W9-CID-451

The Finiteness Obstruction
of C. T. C. Wall

CANADIAN MATHEMATICAL SOCIETY SERIES OF MONOGRAPHS AND ADVANCED TEXTS

Monographies et Études de la Société Mathématique du Canada

EDITORIAL BOARD

Frederick V. Atkinson, Bernhard Banaschewski, Colin W. Clark, Erwin O. Kreyszig (Chairman), and John B. Walsh

Frank H. Clarke   *Optimization and Nonsmooth Analysis*

Erwin Klein and Anthony C. Thompson   *Theory of Correspondences: Including Applications to Mathematical Economics*

I. Gohberg, P. Lancaster, and L. Rodman   *Invariant Subspaces of Matrices with Applications*

Jonathan Borwein and Peter Borwein   *PI and the AGM—A Study in Analytic Number Theory and Computational Complexity*

Subhashis Nag   *The Complex Analytic Theory of Teichmüller Spaces*

Manfred Kracht and Erwin Kreyszig   *Methods of Complex Analysis in Partial Differential Equations with Applications*

Ernst J. Kani and Robert A. Smith   *The Collected Papers of Hans Arnold Heilbronn*

John H. Berglund, Hugo D. Junghenn, and Paul Milne   *Analysis on Semigroups: Function Spaces, Compactifications, Representations*

Victor P. Snaith   *Topological Methods in Galois Representation Theory*

Kalathoor Varadarajan   *The Finiteness Obstruction of C. T. C. Wall*

The Finiteness Obstruction of C. T. C. Wall

KALATHOOR VARADARAJAN

The University of Calgary
Calgary, Alberta, Canada

WILEY

A Wiley-Interscience Publication

JOHN WILEY & SONS

New York • Chichester • Brisbane • Toronto • Singapore

Copyright © 1989 by John Wiley & Sons, Inc.

All rights reserved. Published simultaneously in Canada.

Reproduction or translation of any part of this work
beyond that permitted by Section 107 or 108 of the
1976 United States Copyright Act without the permission
of the copyright owner is unlawful. Requests for
permission or further information should be addressed to
the Permissions Department, John Wiley & Sons, Inc.

Library of Congress Cataloging in Publication Data:

Varadarajan, Kalathoor.
    The finiteness obstruction of C. T. C. Wall/Kalathoor Varadarajan.
       p.   cm.

    "A Wiley-Interscience publication."
    Bibliography: p.
    Includes index.
    1. CW complexes. 2. K-theory. 3. Homotopy theory.   I. Title.
II. Title: Finiteness obstruction of CTC Wall.
QA611.35.V37 1989
514′.223—dc 19   88-30281
                                                 CIP

Printed in the United States of America

10 9 8 7 6 5 4 3 2 1

Math = CB-52K,
Syp   SCIMON

Preface

The material presented in this monograph grew out of seminars I conducted at the University of Calgary for advanced graduate students. We assume that the reader is familiar with basic material in algebraic topology and homotopy theory. Excellent sources of reference are [19] and [26].

In Chapter 1 we give a quick account of the theory of CW-complexes. The importance of CW-complexes in homotopy theory is due to the fact that a map $f\colon X \to Y$ of arcwise-connected CW-complexes is a homotopy equivalence if and only if $f_*\colon \pi_n(X) \to \pi_n(Y)$ is an isomorphism for all $n \geq 1$. We give a complete proof of this celebrated theorem of J. H. C. Whitehead. Another important result of Whitehead asserts that any space homotopically dominated by a CW-complex is itself of the homotopy type of a CW-complex. We also give a complete proof of this result. For this purpose we briefly recall the definition of the geometric realization of a C.S.S. complex as defined by Milnor [12]. Assuming the result that the natural map $\alpha_X\colon |S(X)| \to X$ is a weak homotopy equivalence we complete the proof of Whitehead's result about spaces dominated by CW-complexes. Whitehead used a different geometric realization of the singular semi-simplicial complex of X.

A map $f\colon X \to Y$ will be referred to as a homology equivalence if $f_*\colon H_*(X) \simeq H_*(Y)$ where $H_*(X)$ denotes the singular homology of X with integer coefficients. Any weak equivalence $X \xrightarrow{f} Y$ is a homology equivalence. For simply connected spaces the converse is true. Actually in Chapter 2 the converse will be proved for 0-connected nilpotent spaces. In Chapter 1 we give an example of a homology equivalence $f\colon X \to Y$, further satisfying the condition that $f_*\colon \pi_1(X) \to \pi_1(Y)$ is an isomorphism, but f itself is not a weak homotopy equivalence. We end Chapter 1 with a brief discussion of the method of constructing a Postnikov system for an arcwise-connected space X.

Chapter 2 deals with E. Dror's generalization of Whitehead's theorem for nilpotent spaces. In this chapter and later on we use arguments involving spectral sequences. The reader can refer to Chapter 15 of [4], Chapter 9 of [19], and Chapter 13 of [26] for information about spectral sequences. The motivation for Dror's theorem comes from Stallings' theorem on the relationship between the homology and lower central series of groups [20]. After proving Stallings' theorem we introduce the concept of nilpotent action of a group on another group and then prove Dror's generalization of Whitehead's theorem in its most general form (Theorem 2.8 of Chapter 2). Then we introduce the concept of nilpotent spaces and deal with many examples of nilpotent spaces. For a topological group G that is not connected we obtain necessary and sufficient conditions for the classifying space B_G to be nilpotent, a result originally proved by Roitberg [18]. From the definition of nilpotent spaces, it is evident that Dror's theorem is valid for nilpotent spaces. In fact it is the validity of Dror's theorem combined with successful localization techniques that has made the class of nilpotent CW-complexes important in homotopy theory. For a comprehensive account of localization techniques on the category of nilpotent groups and then on the category of nilpotent spaces the reader could refer to [9].

Chapter 3 is concerned with basic results on the Grothendieck groups of finitely generated projective modules and of all finitely generated modules over a ring R. Letting $\underline{P}(R)$ [resp. $\underline{M}(R)$] denote the full subcategory of finitely generated projective modules (all finitely generated modules) of R-mod, we introduce the unreduced and reduced Grothendieck groups $K_0(R)$ and $\tilde{K}_0(R)$ [resp. $G_0(R)$ and $\tilde{G}_0(R)$] of $\underline{P}(R)$ [resp. $\underline{M}(R)$]. Let $f: R \to S$ be a ring homomorphism. Always there is an induced homomorphism $f_*: K_0(R) \to K_0(S)$. In case S is flat as a right R-module via f we define an induced map $f_*: G_0(R) \to G_0(S)$. We can also regard S as a left R-module via f. When $S \in \underline{P}(R)$ [resp. $\underline{M}(R)$] we introduce "restriction of scalars" $f^*: K_0(S) \to K_0(R)$ [resp. $f^*: \tilde{G}_0(S) \to G_0(R)$] and we determine $K_0(R)$, $\tilde{K}_0(R)$, $G_0(R)$, and $\tilde{G}_0(R)$ in some special cases. We introduce the Cartan maps $c: K_0(R) \to G_0(R)$ and $\tilde{K}_0(R) \to \tilde{G}_0(R)$. We then study the class of homologically regular rings and prove that the Cartan maps $c: K_0(R) \to G_0(R)$ and $c: \tilde{K}_0(R) \to \tilde{G}_0(R)$ are isomorphisms when R is homologically regular. The material presented here is contained in Bass's book *Algebraic K-Theory* [1], but is scattered all over the book. Collecting the results needed for our purposes will help the reader immensely and Chapter 3 exactly serves this purpose.

In Chapter 4 we first develop the theory of Dedekind domains in a somewhat classical spirit. After dealing with concepts such as integral elements and integral closure, we introduce a Dedekind domain as an integral domain that is noetherian, integrally closed, and having the property that every nonzero prime ideal is maximal. Next we prove that the integers in any algebraic number field form a Dedekind domain. Then we introduce the concept of fractional ideals and through them the notion of the ideal class group. Then we prove that every nonzero ideal in a Dedekind domain can be expressed as a product of powers of maximal ideals in an essentially unique way. Then we go on to prove Steinitz's theorem and then establish the fundamental result that the reduced Grothendieck group $\tilde{K}_0(R)$ of a Dedekind domain R is isomorphic to the class group $Cl(R)$ of R. Actually Chapter 4 is an essential part of algebraic number theory. A part of the results here, like Steinitz's theorem and the isomorphism between $\tilde{K}_0(R)$ and $Cl(R)$, are briefly dealt with in Milnor's book on algebraic K-theory. But the emphasis there is on the higher K-groups K_1 and K_2 rather than $\tilde{K}_0$. The isomorphism of

$\tilde{K}_0(R)$ with $Cl(R)$ is very crucial in our study of Wall's finiteness obstruction and justifies inclusion of these results in the present monograph.

In Chapter 5, our main goal is to prove Dock Sang Rim's theorem asserting that for a cyclic group π of order p, where p is a prime, $\tilde{K}_0(Z\pi) \sim \tilde{K}_0(Z[\omega])$ $\simeq Cl(Z[\omega])$ where $\omega = \exp(2\pi\sqrt{-1}/p)$. We introduce for any *finite* group π and any left $Z\pi$-module A, the complete cohomology $\hat{H}(\pi, A)$ of π with coefficients in A. For any exact sequence $0 \to A' \to A \to A'' \to 0$ of $Z\pi$-modules we obtain an exact sequence

$$\cdots \to \hat{H}^n(\pi, A') \to \hat{H}^n(\pi, A) \to \hat{H}^n(\pi, A'') \to \hat{H}^{n+1}(\pi, A) \to$$

of complete cohomology groups for $n \in Z$. For any commutative ring K we introduce notions of weakly projective and weakly injective $K\pi$-modules. It turns out that these two apparently different notions are the same and are equivalent to the vanishing of the complete cohomology $\hat{H}(\pi, \text{Hom}_K(A, A))$. We then try to specialize to the case when π is a p group and A is a $Z_p\pi$-module, where $Z_p = Z/pZ$. In this case it turns out that A is weakly projective as a $Z\pi$-module $\Leftrightarrow$ it is weakly projective as a $Z_p\pi$-module $\Leftrightarrow$ it is free as a $Z_p\pi$-module.

We then introduce the class of cohomologically trivial π-modules for any finite group π and characterize these as modules with finite projective dimension over $Z\pi$ or equivalently those with finite injective dimension over $Z\pi$. Surprisingly, these also turn out to be the same as the modules with projective dimension less than or equal to 1, equivalently with injective dimension less than or equal to 1. These results from the cohomology theory of finite groups are used in the proof of Dock Song Rim's theorem. As indicated in the text, an alternative source for results on complete cohomology of finite groups is [4].

In Chapter 6 we give an account of C. T. C. Wall's work on finiteness conditions on CW-complexes. As indicated in Chapter 1 of this monograph, one of the results of J. H. C. Whitehead asserted that any space homotopically dominated by a CW-complex is itself of the homotopy type of a CW-complex. J. H. C. Whitehead raised the following question: If X is dominated by a finite complex, is it of the homotopy type of a finite complex [27]? He considered this to be an extremely difficult problem. When X is simply connected Milnor had an affirmative answer to this question but this result remains unpublished. It is in the nonsimply connected case that the solution to the problem really needed new ideas involving algebraic K-theory. These were supplied by C. T. C. Wall in his paper "Finiteness conditions for CW-complexes" [24]. C. T. C. Wall's method also makes repeated use of a construction owing to Milnor. This construction itself appeared in print for the first time in Wall's paper. While we give a complete account of C. T. C. Wall's results, our development differs from his at one crucial place. His proofs of certain results (for instance Propositions 1.5 and 1.6, and Lemma 2.9 in Chapter 6 of the present monograph) heavily depended on Whitehead's results on simple homotopy types. Our proofs are algebraic in nature. In Section 1, conditions F_n of C. T. C. Wall are introduced and their equivalence to X being homotopically equivalent to a CW-complex with finite n skeleton or X being dominated by a

CW-complex with finite n skeleton is established. From these one immediately obtains necessary and sufficient conditions for X to be homotopically equivalent to a complex K with finite n skeleton for each n. In Section 2 necessary and sufficient conditions are obtained for X to be homotopically equivalent to a complex of finite dimension. Combining the results of Section 1 with these, the finiteness obstruction $\tilde{\omega}(X) \in \tilde{K}_0(Z\pi)$ where $\pi = \pi_1(X)$ is introduced and the result that X is homotopically equivalent to a finite complex if and only if $\tilde{\omega}(X) = 0$ is proved. Also given any finitely presented group π and any element $x \in \tilde{K}_0(Z\pi)$, a method is given for constructing a finitely dominated space X with $\pi_1(X) \simeq \pi$ and $\tilde{\omega}(X) = x$. Using the fact that there exist finitely presented groups π with $\tilde{K}_0(Z\pi) \neq 0$, one shows that Whitehead's problem in general has a negative answer. In Section 3 we give Gersten's proof of the product formula for the Wall obstruction.

Chapter 7 is devoted to the study of finitely dominated nilpotent spaces. The results here are mainly due to Mislin. One of the main results of Mislin gives a completely homological criterion for a nilpotent space to be finitely dominated. A connected space X is said to be quasi-finite if $H_i(X)$ is finitely generated for all i and there exists an integer l with $H_i(X) = 0$ for $i > l$. Mislin's result asserts that a nilpotent space X is finitely dominated if and only if it is quasi-finite. We give a complete proof of this in Section 1. Also it turns out that a finitely dominated nilpotent space with infinite fundamental group is of the homotopy type of a finite complex. In Section 2 we study restrictions that the Wall obstruction $\tilde{\omega}(X)$ of a finitely dominated nilpotent space should satisfy if π is a finite p group or a finite abelian group. Results from algebraic number theory and induction theorems in K-theory play a crucial role in these proofs.

I would like to express my sincere gratitude to the University of Calgary for the award of a Killam Residential Fellowship which enabled me to complete my work on this monograph. My sincere thanks also go to Dr. Piccinini for suggesting that I publish this monograph in the *Canadian Mathematical Society Series*.

In a monograph of this type there is bound to be some overlap with some books that have appeared earlier. Chapters 2, 6, and 7 will be appearing in a book for the *first* time. Though part of the material in Chapters 3–5 can be found in other books the main trend in those other books is completely different and they cannot all be found in the same book. Collecting these in one source helps the reader immensely. Also these are very crucial for the development of the material in this monograph. This justifies their inclusion in this monograph.

Finally I would like to thank Gisele Vezina for her excellent job of typing.

KALATHOOR VARADARAJAN

Calgary, Alberta, Canada

Contents

The Finiteness Obstruction of C. T. C. Wall

Chapter One

CW-Complexes and J. H. C. Whitehead's Theorems

Our main aim in this chapter is to give *brief* indications of proofs of some basic results of J. H. C. Whitehead in homotopy theory that we need for our later developments. In addition to the original papers of Whitehead [27, 28] there are many sources in literature giving a fairly thorough account of the theory of CW-complexes and all the basic results of Whitehead concerning their homotopy theory [11, 19, 26]. Since the present book is about the "Wall" obstruction of finitely dominated spaces, we will assume familiarity on the part of the reader with basic material in homotopy theory. The interested reader may refer to the abovementioned sources for a comprehensive account of the theory of CW-complexes.

1 WEAK HOMOTOPY EQUIVALENCES AND HOMOLOGY EQUIVALENCES

For any integer $k \geq 1$ let $\mathbb{R}^k$ denote the real Euclidean space of dimension k. $E^k = \{(x_1, \ldots, x_k) \in \mathbb{R}^k | \Sigma_{i=1}^k x_i^2 \leq 1\}$ is the unit disk in $\mathbb{R}^k$ and $S^{k-1} = \{(x_1, \ldots, x_k) \in \mathbb{R}^k | \Sigma_{i=1}^k x_i^2 = 1\}$ is the unit sphere in $\mathbb{R}^k$. We define E^0 to be a point and S^{-1} to be the empty set. Let $p_0 = (1, 0, \ldots, 0) \in S^{k-1}$. In this section n will be an integer greater than or equal to 0 and (X, A) will denote a pair of topological spaces with $A \neq \varnothing$.

Definition 1.1

We say that the pair (X, A) is *n-connected* if for $0 \leq k \leq n$, every map $\alpha: (E^k, S^{k-1}) \to (X, A)$ is homotopic *relative to* S^{k-1} to a map $\beta: (E^k, S^{k-1}) \to (X, A)$ with $\beta(E^k) \subset A$.

1

Observe that for $k = 0$, this is equivalent to the condition that every point in X can be joined by means of an arc in X to some point in A. Equivalently this means that each arcwise-connected component of X intersects A.

The following is a standard result in homotopy theory.

Proposition 1.2

Let $x_0 \in A$, k an integer greater than or equal to 1, and $\alpha: (E^k, S^{k-1}, p_0) \to (X, A, x_0)$ a given map. Then $[\alpha] = 0$ in $\pi_k(X, A, x_0)$ if and only if α is homotopic relative to S^{k-1} to a map $\beta: (E^k, S^{k-1}, p_0) \to (X, A, x_0)$ with $\beta(E^k) \subset A$.

For a proof see Theorem 1, Chapter 7, Section 2 of [19].

Corollary 1.3

(X, A) is n-connected if and only if each arc component of X intersects A and $\pi_k(X, A, a) = 0$ for each $a \in A$ and $1 \leq k \leq n$.

Remark 1.4

Suppose $A = \{x_0\}$. Then (X, A) is 0-connected if and only if X is arcwise-connected. Also if $n \geq 1$, then (X, A) is n-connected if and only if X is arcwise-connected and $\pi_k(X, x_0) = 0$ for $1 \leq k \leq n$. However when X is arcwise-connected, since $\pi_k(X, a) \simeq \pi_k(X, b)$ for any two points a, b in X we may write the preceding condition as $\pi_k(X) = 0$ for $1 \leq k \leq n$. Thus we may formulate the following.

Definition 1.5

Let X be a nonempty topological space. Then X is said to be *n-connected* if it is arcwise-connected and $\pi_k(X) = 0$ for $1 \leq k \leq n$.

For any $x_0 \in X$, there is a standard action of $\pi_1(X, x_0)$ on $\pi_n(X, x_0)$ for all $n \geq 1$. Under this action $\pi_1(X, x_0)$ acts on itself by inner automorphisms. For any pair (X, A) with $A \neq \varnothing$ and any $x_0 \in A$ there is a standard action of $\pi_1(A, x_0)$ on $\pi_{n+1}(X, A, x_0)$. We choose a base point $x_0 \in A$ and fix it throughout the following discussion. Because of this we will not write the base point while writing the homotopy groups. Let $\omega_n(X)$ [resp. $\omega_{n+1}(X, A)$] be the subgroup of $\pi_n(X)$ [resp. $\pi_{n+1}(X, A)$] generated by $\omega_\#(\alpha) \cdot \alpha^{-1}$ with $\omega \in \pi_1(X)$ and $\alpha \in \pi_n(X)$ [resp. $\omega \in \pi_1(A)$ and $\alpha \in \pi_{n+1}(X, A)$]. Clearly $\omega_1(X) = [\pi_1(X), \pi_1(X)]$ is the commutator subgroup of $\pi_1(X)$. It is known that $\omega_2(X, A)$ is a normal subgroup of $\pi_2(X, A)$ with an abelian quotient. For $n \geq 2$, the groups $\pi_n(X)$ and $\pi_{n+1}(X, A)$ are abelian. Let $\pi_n^*(X) = \pi_n(X)/\omega_n(X)$ [resp. $\pi_{n+1}^*(X, A) = \pi_{n+1}(X, A)/\omega_{n+1}(X, A)$]. It is known that the Hurewicz homomorphisms $\rho: \pi_n(X) \to H_n(X)$ and $\rho: \pi_{n+1}(X, A) \to H_{n+1}(X, A)$ carry $\omega_n(X)$ [resp. $\omega_{n+1}(X, A)$] to 0, thus yielding homomorphisms

$$\bar{\rho}: \pi_n^*(X) \to H_n(X),$$

$$\bar{\rho}: \pi_{n+1}^*(X, A) \to H_{n+1}(X, A),$$

for all $n \geq 1$. With these notations we have:

Theorem 1.6 (Absolute Hurewicz Theorem)

Let X be an $(n - 1)$-connected space with $n \geq 1$. Then $\bar{\rho}: \pi_n^*(X) \to H_n(X)$ is an isomorphism.

Theorem 1.7 (Relative Hurewicz Theorem)

Let A be path-connected and (X, A) be $(n - 1)$-connected with $n \geq 2$. Then $\bar{\rho}: \pi_n^*(X, A) \to H_n(X, A)$ is an isomorphism.

For a proof see either Chapter 7, Section 5 of [18] or Chapter 4, Section 7 of [26].

The following are easy corollaries of Theorems 1.6 and 1.7.

Corollary 1.8

(i) *Let X be $(n - 1)$-connected with $n \geq 2$. Then $H_i(X) = 0$ for $1 \leq i < n$ and $\rho: \pi_n(X) \simeq H_n(X)$.*

(ii) *Suppose X is 1-connected and $H_i(X) = 0$ for $1 \leq i < n$ ($n \geq 2$ clearly). Then $\pi_i(X) = 0$ for $1 \leq i < n$ and $\rho: \pi_n(X) \simeq H_n(X)$.*

(iii) *Let (X, A) be $(n - 1)$-connected with $n \geq 2$ and A be arc-connected. Then $H_i(X, A) = 0$ for $i < n$ and $\bar{\rho}: \pi_n^*(X, A) \simeq H_n(X, A)$.*

(iv) *Let (X, A) be $(n - 1)$-connected with $n \geq 2$ and A be simply connected. Then $H_i(X, A) = 0$ for $i < n$ and $\rho: \pi_n(X, A) \simeq H_n(X, A)$.*

(v) *Let (X, A) be 1-connected and A also be 1-connected. Suppose $H_i(X, A) = 0$ for $i < n$ ($n \geq 2$ clearly). Then (X, A) is $(n - 1)$-connected and $\rho: \pi_n(X, A) \simeq H_n(X, A)$.*

Given any map $f: X \to Y$ let M_f denote the mapping cylinder of f.

Definition 1.9

We say that f *is n-connected* if and only if the pair (M_f, X) is n-connected.

Lemma 1.10

Let X, Y be 0-connected spaces, $x_0 \in X$ and $y_0 \in Y$. Let $f: (X, x_0) \to (Y, y_0)$. Then f is n-connected if and only if $f_*: \pi_k(X, x_0) \to \pi_k(Y, y_0)$ is an isomorphism for $k < n$ and onto for $k = n$.

Proof. Immediate consequence of the homotopy exact sequence of (M_f, X) and Corollary 1.3. $\square$

Theorem 1.11 (J. H. C. Whitehead)

Let $f: X \to Y$ be a map of 0-connected spaces.

(i) If f is n-connected, then $f_*: H_k(X) \to H_k(Y)$ is an isomorphism for $k < n$ and onto for $k = n$.

(ii) Conversely, suppose X and Y are 1-connected and $f_*: H_k(X) \to H_k(Y)$ is an isomorphism for $k < n$ and onto for $k = n$. Then f is n-connected.

Proof. The condition $H_k(M_f, X) = 0$ for $k \leq n$ is clearly equivalent to stating that $f_*: H_k(X) \to H_k(Y)$ is an isomorphism for $k < n$ and onto for $k = n$. Now (i) is an immediate consequence of Corollary 1.8(iii) and (ii) is an immediate consequence of Corollary 1.8(v). $\square$

Definition 1.12

A map $f: X \to Y$ is called a *weak homotopy equivalence* if it is n-connected for all $n \geq 0$.

From Lemma 1.10 we see that a map $f: X \to Y$ of 0-connected spaces is a weak homotopy equivalence if and only if $f_*: \pi_k(X, x_0) \simeq \pi_k(Y, f(x_0))$ for all k.

Definition 1.13

A map $f: X \to Y$ is referred to as a *homology equivalence* if $f_*: H_k(X) \to H_k(Y)$ is an isomorphism for all $k \geq 0$.

The following is an immediate consequence of Theorem 1.11.

Theorem 1.14 (J. H. C. Whitehead)

Let $f: X \to Y$ be a map of 0-connected spaces.

(i) If f is a weak homotopy equivalence, then f is a homology equivalence.

(ii) Conversely, if X and Y are both 1-connected and f a homology equivalence, then f is a weak homotopy equivalence.

Remark 1.15

Later on, in Chapter 2 we will prove E. Dror's result which asserts that the converse Theorem 1.14(ii) is valid even when X and Y are nilpotent spaces.

For any space X let $\pi_0(X)$ denote the set of arc components of X. Given a map $f: X \to Y$ there is an induced function $f_*: \pi_0(X) \to \pi_0(Y)$ defined as follows.

If A is any arcwise-connected component of X, then $f_*(A)$ equals the arc component of Y containing $f(A)$. With this understanding, Lemma 1.10 can be extended as follows.

Lemma 1.16

A map $f: X \to Y$ is n-connected if and only if $f_*: \pi_0(X) \to \pi_0(Y)$ is a bijection and $f_*: \pi_k(X, x) \to \pi_k(Y, f(x))$ is an isomorphism for $1 \le k < n$ and onto for $k = n$, whatever $x \in X$ is.

2 CW-COMPLEXES

Let X, Y be topological spaces, A a closed subspace of X, and $f: A \to Y$ a map. The adjunction space $Y \cup_f X$ is defined to be the quotient space obtained from the topological union $Y \cup X$ by identifying $a \in A$ with $f(a)$ in Y for each $a \in A$. Let $\eta: Y \cup X \to Y \cup_f X$ denote the quotient map. It is clear that $\eta|Y$ is a homeomorphism of Y onto the closed subspace $\eta(Y)$ of $Y \cup_f X$ and that $\eta|(X - A)$ is a homeomorphism of $X - A$ onto the open subset $\eta(X - A)$ of $Y \cup_f X$. Moreover $Y \cup_f X = \eta(Y) \cup \eta(X - A)$ (disjoint union). If $A = \varnothing$, the adjunction space is just the topological union of Y and X. We are particularly interested in the following situation. Let E_λ be a disk of dimension n_λ for each $\lambda \in J$ and $\dot{E}_\lambda = S_\lambda^{n_\lambda - 1}$ the boundary of E_λ. Let $f_\lambda: S_\lambda^{n_\lambda - 1} \to Y$ be given maps. Take $X = \bigcup_{\lambda \in J} E_\lambda^{n_\lambda}$, $A = \bigcup_{\lambda \in J} S_\lambda^{n_\lambda - 1}$ (topological unions), and $f: A \to Y$ to be the map given by $f|S_\lambda^{n_\lambda - 1} = f_\lambda$. The adjunction space $Y \cup_f X$ in this particular case is said to be obtained from Y by attaching cells E_λ by means of the maps $f_\lambda: S_\lambda^{n_\lambda - 1} \to Y$. The quotient map $\eta: Y \cup X \to Y \cup_f X$ has the property that $\eta|E_\lambda$ is an extension of $f_\lambda: S_\lambda^{n_\lambda - 1} \to Y$. If $e_\lambda = \eta(E_\lambda)$ and $\dot{e}_\lambda = f_\lambda(\dot{E}_\lambda)$, it is customary to call e_λ the closed cells of $Y \cup_f X$, $\dot{e}_\lambda$ the boundary of e_λ. If $g_\lambda = \eta|E_\lambda$, then $g_\lambda: (E_\lambda, \dot{E}_\lambda) \to (e_\lambda, \dot{e}_\lambda)$ is a relative homeomorphism—namely $g_\lambda|(E_\lambda - \dot{E}_\lambda)$ is a homeomorphism of $E_\lambda - \dot{E}_\lambda$ onto $e_\lambda - \dot{e}_\lambda$. It is customary to denote $Y \cup_f X$ in this case by $Y \cup_{f_\lambda} \bigcup e_\lambda$. If Y is Hausdorff, it can be shown that $Y \cup_{f_\lambda} (\bigcup e_\lambda)$ is also Hausdorff. The map g_λ will be called the characteristic map of the cell e_λ.

Definition 2.1

A *CW-complex* is a topological space X together with a sequence $X_0 \subset X_1 \subset X_2 \subset \cdots$ of *closed* subspaces of X satisfying the following conditions:

(i) $X = \bigcup_{n \geq 0} X_n$.

(ii) X_0 is discrete.

(iii) For each $n \geq 1$, X_n is obtained from X_{n-1} by attaching a family of n-dimensional cells.

(iv) The topology of X is the "coherent topology" determined by the X_n's. This means a subset C of X is closed in X if and only if $C \cap X_n$ is closed in X_n for each $n \geq 0$.

Let $X_n = X_{n-1} \cup_{f_\alpha} (\bigcup e_\alpha^n)_{\alpha \in J_n}$ and $g_\alpha : (E_\alpha^n, \dot{E}_\alpha^n) \to (e_\alpha^n, \dot{e}_\alpha^n)$ be the characteristic map for e_α^n. For giving a CW-structure on X we have to specify the characteristic maps for the various cells. We may define $X_{-1} = \varnothing$ and regard X_0 as obtained from X_{-1} by attaching 0-dimensional cells. Without proofs we mention the following properties of a CW-complex X.

(2.2) X is Hausdorff (in fact paracompact).

(2.3) If $\{e_\alpha^n\}_{\alpha \in J_n}$, $n \geq 0$, denote the "closed" cells of X, then a subset C of X is closed in X if and only if $C \cap e_\alpha^n$ is closed in e_α^n for each e_α^n and each $n \geq 0$. This property is otherwise stated as follows: The topology of X is the "weak topology" determined by the cells of X.

Let X be a Hausdorff space. Suppose there exist continuous onto maps $f_\alpha^n : E_\alpha^n \to e_\alpha^n$ with $e_\alpha^n \subset X$, $\alpha \in J_n$ and $n \geq 0$, satisfying the following conditions:

1. Denoting $f_\alpha(S_\alpha^{n-1})$ by $\dot{e}_\alpha^n$, we require $f_\alpha : (E_\alpha^n, \dot{E}_\alpha^n) \to (e_\alpha^n, \dot{e}_\alpha^n)$ to be a relative homeomorphism.

2. If $X_n = \bigcup_{\alpha \in J_k, \; 0 \leq k \leq n} e_\alpha^k$, then $\dot{e}_\alpha^n \subset X_{n-1}$ (in case $n = 0$, we set $X_{-1} = \varnothing$).

3. Writing $\text{Int } e_\alpha^n = e_\alpha^n - \dot{e}_\alpha^n$ we require that $X = \bigcup_{\alpha \in J_n, \; n \geq 0} \text{Int } e_\alpha^n$, a disjoint union.

In this case X is referred to as a cell complex. The e_α^n's provide the cell structure on X. We refer to the e_α^n's as the cells of X.

Remarks 2.4

The notation Int e_α^n is somewhat misleading. Int e_α^n are not necessarily open in X. A cell complex in general does not satisfy conditions (ii) and (iv) in the definition of a CW-complex. Also the X_n's need not be closed subsets of X.

A *subcomplex* Y of X is a cell complex such that each (closed) cell of Y is a cell of X. A cell complex is said to be finite (resp. countable) if the number of cells in it is finite (resp. countable).

We state again without proofs the following properties of CW-complexes.

(2.5) Any subcomplex of a CW-complex is itself a CW-complex.

(2.6) Given any cell e of a CW-complex X there exists a finite subcomplex K of X with $e \subset K$. This property is known as closure finiteness of X.

A CW-complex can alternatively be defined as a cell complex satisfying (2.3) and (2.6).

(2.7) If A is a subcomplex of a CW-complex X, then the inclusion $A \to X$ is a cofibration.

Definition 2.8

By a *CW-pair* (X, A) we mean a CW-complex X and a subcomplex A of X.

We will say that $\dim(X, A) \leq n$ if any cell e of X that is not a cell of A satisfies $\dim e \leq n$.

Remark 2.9

There is a more general notion of a relative CW-complex (X, Y). The interested reader may refer to [11, 26].

If X, Y are CW-complexes, on $X \times Y$ we define a cell complex structure by taking the products $e \times e'$ as cells of $X \times Y$ where e is a cell of X, e' a cell of Y. We put the topology coherent with the cells $e \times e'$ on $X \times Y$. This in general will be finer than the product topology.

Lemma 2.10

Let $X = A \cup_{f_\alpha} \bigcup \{e_\alpha^n\}_{\alpha \in J}$ with A Hausdorff. Let $h_\alpha : (E_\alpha^n, \dot{E}_\alpha^n)$ be the characteristic map for e_α^n. Let (Y, B) be any pair of spaces and $g : (X, A) \to (Y, B)$ be a given map. Suppose for each $\alpha \in J$, $H_\alpha : (E_\alpha^n \times I, \dot{E}_\alpha^n \times I) \to (Y, B)$ is a *homotopy of* $g|e_\alpha^n \circ h_\alpha$ *relative to* $\dot{E}_\alpha^n$. *Then there exists a homotopy* $H:(X, A) \times I \to (Y, B)$ *of* g *relative to* A *satisfying* $(H|e_\alpha^n \times I) \circ (h_\alpha \times \mathrm{Id}_I) = H_\alpha$ *for each* $\alpha \in J$.

Proof. Let $\eta: A \cup \bigcup_{\alpha \in J} E_\alpha^n \to X$ denote the quotient map. Then it is easy to check that $(A \cup \bigcup_{\alpha \in J} E_\alpha^n) \times I \xrightarrow{\eta \times \mathrm{Id}} X \times I$ is a quotient map. Let

$$G: \left(A \cup \bigcup_{\alpha \in I} E_\alpha^n \right) \times I \to Y$$

be given by

$$G:(a, t) = g(a) \qquad \forall\, a \in A,$$

$$G|E_\alpha^n \times I = H_\alpha.$$

For any $u \in \dot{E}_\alpha^n$ we have $G(u, t) = g \circ h_\alpha(u)$. Hence G induces a map $H: X \times I \to Y$ satisfying $H \circ (\eta \times \mathrm{Id}) = G$. It is clear that H satisfies the requirements of the lemma. $\square$

Proposition 2.11

Let (Y, B) *be* n-*connected and* $f:(X, A) \to (Y, B)$ *be a map with* (X, A) *a* CW-*pair satisfying* $\dim(X, A) \le n$. *Then* f *is homotopic relative to* A *to a map* $\varphi:(X, A) \to (Y, B)$ *with* $\varphi(X) \subset B$.

Proof. For every integer $q \ge 0$ let $X^q = A \cup$ (cells of dimension $\le q$ of X) and $X^{-1} = A$. Then X^q is obtained from X^{q-1} by adjoining a family of q cells and $X^n = X$. By induction on q we will prove that f is homotopic relative to A to a map $g_q:(X, A) \to (Y, B)$ satisfying $g_q(X^q) \subset B$. For $q = -1$ this is clearly satisfied. We can take $g_{-1} = f$. Assume $0 \le q \le n$ and that we have a map $g_{q-1}:(X, A) \to (Y, B)$ homotopic to f relative to A and satisfying $g_{q-1}(X^{q-1}) \subset B$. Let $h_\alpha:(E_\alpha^q, \dot{E}_\alpha^q) \to (e_\alpha^q, \dot{e}_\alpha^q)$ be the characteristic maps of the q cells of X not in A. Since $g_{q-1}(X^{q-1}) \subset B$, we see that $g_{q-1} \circ h_\alpha(\dot{E}_\alpha^q) \subset B$. Since $q \le n$ and (Y, B) is n-connected, we see that $g_{q-1} \circ h_\alpha:(E_\alpha^q, \dot{E}_\alpha^q) \to (Y, B)$ is homotopic relative to $\dot{E}_\alpha^q$ to a map carrying E_α^q into B. From Lemma 2.10, we see that $g_{q-1}:(X^q, X^{q-1}) \to (X, B)$ is homotopic relative to X^{q-1} to a map $\theta:(X^q, X^{q-1}) \to (Y, B)$ satisfying $\theta(X^q) \subset B$. Since $X^q \to X$ is a cofibration, the homotopy between $g_{q-1}|X^q$ and θ can be extended to a homotopy of g_{q-1} to an extension $g_q: X \to Y$ of θ. Then $f \sim g_q(\text{rel. } A)$ and $g_q(X^q) = \theta(X^q) \subset B$. This completes the inductive step. $\varphi = g_n$ satisfies the requirements of the proposition. $\square$

Proposition 2.12

Let (Y, B) be n-connected for all $n \geq 0$. Let (X, A) be any CW-pair. Then any map $f: (X, A) \to (Y, B)$ is homotopic relative to A to a map $g: (X, A) \to (X, B)$ satisfying $g(X) \subset B$.

Proof. Let $X^q = A \cup$ cells of X of dimension $\leq q$. From Proposition 2.11 and the fact that $X^q \to X$ is a cofibration we get a sequence of maps

$$H_k: (X, A) \times I \to (Y, B) \quad \text{for } k \geq 0$$

satisfying the following conditions:

(i) $H_0(x, 0) = f(x) \; \forall \; x \in x$.
(ii) $H_k(x, 1) = H_{k+1}(x, 0) \; \forall \; x \in x$.
(iii) H_k is a homotopy relative to X^{k-1}.
(iv) $H_k(X^k \times 1) \subset B$.

Then $H: (X, A) \times I \to (Y, B)$ defined by

$$H(x, t) = H_{k-1}\left(x, [t - (1 - 1/k)]/[1/k - 1/(k+1)]\right)$$

for $1 - 1/k \leq t \leq 1 - 1/(k+1)$ $(k \geq 1)$ is easily seen to be a homotopy relative to A to a map carrying X to B.

In fact for $x \in X^k$ we get $H(x, t) \in B$ whenever $t \geq 1 - 1/(k+2)$. $\square$

Definition 2.13

When (Y, B) is n-connected for all n, we will say that (Y, B) is ∞-connected. A map $f: X \to Y$ will be said to be ∞-connected if it is n-connected for all n.

Theorem 2.14

Let $f: X \to Y$ be n-connected (n an integer greater than or equal to 0 or $n = \infty$). Let (P, Q) be a CW-pair with $\dim(P, Q) \leq n$ [when $n = \infty$, this just means that (P, Q) is a CW-pair]. Let $h: P \to Y$, $g: Q \to X$ be maps satisfying $h|Q = f \circ g$. Then there exists a map $g': P \to X$ such that $g'|Q = g$ and $f \circ g' \sim h$ rel. Q.

Proof. Let M_f be the mapping cylinder of f. Let $i: X \to M_f$ and $j: Y \to M_f$ denote the canonical inclusions and $r: M_f \to Y$ the canonical retraction. More explicitly $i(x) = \langle x, 0 \rangle$, $j(y) = \langle y \rangle$ and $r\langle x, t \rangle = f(x)$, $r\langle y \rangle = y$ for any $x \in X$, $y \in Y$. Here the equivalence class of an element u is indicated by $\langle u \rangle$. We know that j is a homotopy equivalence with r its homotopy inverse. Also

$$
\begin{array}{ccc}
X & \xrightarrow{\ i\ } & M_f \\
\| & & \downarrow{r} \\
X & \xrightarrow[\ f\ ]{} & Y
\end{array}
$$

is clearly commutative. Consider $G: Q \times I \to M_f$ defined by $G(g, t) = \langle g(q), t \rangle$. Then $G(q, 0) = \langle g(q), 0 \rangle = i \circ g(q)$ and $G(q, 1) = \langle g(q), 1 \rangle = \langle f \circ g(q) \rangle = j \circ h(q)$. Moreover $rG(q, t) = f \circ g(q) = h(q)$ for all $t \in I$. Thus G is a homotopy between $i \circ g$ and $j \circ (h|Q)$ with the additional property that $r \circ G(q, t) = h(q)$ for all $q \in Q$ and $t \in I$. $j \circ h: P \to M_f$ is an extension of $j \circ (h|Q)$. Since $Q \to P$ is a cofibration, there exists a map $F: P \times I \to M_f$ with $F|Q \times I = G$ and $F(p, 1) = j \circ h(p)$ for all $p \in P$. Let $h': P \to M_f$ be given by $h'(p) = F(p, 0)$. Then $h'|Q = i \circ g$. Moreover $r \circ F: P \times I \to Y$ is a homotopy between $r \circ h'$ and $r \circ j \circ h$ relative to Q. Since $h'(Q) \subset X$ in M_f we can regard h' as a map of (P, Q) into (M_f, X). From Proposition 2.11 (or Proposition 2.12 in case $n = \infty$) we see that h' is homotopic relative to Q to a map $g': (P, Q) \to (M_f, X)$ with $g'(P) \subset X$. We regard g' as a map of P in X. Then $g'|Q = h'|Q = g$. Also

$$
f \circ g' = r \circ i \circ g' \underset{\text{rel. } Q}{\sim} r \circ h' \underset{\text{rel. } Q}{\sim} r \circ j \circ h = h.
$$

This completes the proof of Theorem 2.14. $\square$

Corollary 2.15

Let $f: X \to Y$ be an n-connected map (n finite or infinite). Consider the map $f_\#: [P, X] \to [P, Y]$ (where $[P, X]$ denotes the set of homotopy classes of maps of P into X).

(i) If P is a CW-complex of dimension less than or equal to n, then $f_\#$ is surjective.

(ii) If P is a CW-complex of dimension less than or equal to $n - 1$, then $f_\#$ is bijective.

Proof. Take $Q = \varnothing$ in Theorem 2.14. Then we see that given any $h: P \to Y$ there exists a $g': P \to X$ with $f \circ g' \sim h$. This proves surjectivity of $f_\#: [P, X] \to [P, Y]$ whenever $\dim P \le n$.

To prove (ii) we have only to show that if $\dim P \le n - 1$, then $f_\#: [P, X] \to [P, Y]$ is injective. In this case we will apply Theorem 2.14 to the CW-pair $(P \times I, P \times \dot{I})$. Let g_0, g_1 be maps $P \to X$ with $f \circ g_0 \sim f \circ g_1$. Let

$H: P \times I \to Y$ be given by $g(p, 0) = g_0(p)$ and $g(p, 1) = g_1(p)$ for all $p \in P$. Then $H|P \times \dot{I} = f \circ g$. Hence by Theorem 2.14 there exists a map $G: P \times I \to X$ such that $G|P \times \dot{I} = g$. This shows that $g_0 \sim g_1$. Hence $f_{\#}$ is injective. $\square$

Theorem 2.16 (J. H. C. Whitehead)

A map $f: X \to Y$ between CW-complexes is a homotopy equivalence if and only if it is a weak homotopy equivalence.

Proof. We have only to prove that a weak homotopy equivalence between CW-complexes is actually a homotopy equivalence. Let $f: X \to Y$ be a weak homotopy equivalence between CW-complexes. From Corollary 2.15 we see that $f_{\#}: [X, X] \to [X, Y]$ and $f_{\#}: [Y, X] \to [Y, Y]$ are bijections. From the latter we see that there exists a map $g: Y \to X$ with $f \circ g \sim \text{Id}_Y$. Now, $f_{\#}[g \circ f] = [f \circ g \circ f] = [\text{Id}_Y \circ f] = [f] = f_{\#}[\text{Id}_X]$ in $[X, Y]$. From the former bijection we see that $g \circ f \sim \text{Id}_X$. Thus f is a homotopy equivalence with g as its homotopy inverse. $\square$

3 SPACES DOMINATED BY CW-COMPLEXES

Definition 3.1

A space X is said to be *dominated* by another space Y in homotopy if there exist maps $f: X \to Y$, $g: Y \to X$ with $g \circ f \sim \text{Id}_X$.

In this section we will be giving the proof of another result of Whitehead that asserts that any space dominated by a CW-complex is itself of the homotopy type of a CW-complex. The proof will make use of the geometric realization of a semi-simplicial complex. For any integer $n \geq 0$ the standard n simplex Δ_n in $\mathbb{R}^{n+1}$ is the subspace $\{(x_0, x_1, \ldots, x_n) \in \mathbb{R}^{n+1}|x_i \geq 0, \sum_{i=0}^{n} x_i = 1\}$ of $\mathbb{R}^{n+1}$. The face and degeneracy maps

$$\partial_i: \Delta_n \to \Delta_{n+1} \quad \text{for } 0 \leq i \leq n + 1,$$

$$s_i: \Delta_{n+1} \to \Delta_n \quad \text{for } 0 \leq i \leq n,$$

are given by

$$\partial_i(x_0, \ldots, x_n) = (x_0, \ldots, x_{i-1}, 0, x_i, \ldots, x_n),$$

$$s_i(u_0, \ldots, u_{n+1}) = (u_0, \ldots, u_{i-1}, u_i + u_{i+1}, u_{i+2}, \ldots, u_{n+1}).$$

The following identities are easily checked.

$$\partial_j \partial_i = \partial_i \partial_{j-1} \quad \text{for } i < j,$$

$$s_j \partial_i = \partial_i s_{j-1} \quad \text{for } i < j,$$

$$s_j \partial_i = \text{Id} \quad \text{if } i = j \text{ or } j + 1,$$

$$s_j \partial_i = \partial_{i-1} s_j \quad \text{if } i > j + 1,$$

$$s_j s_i = s_i s_{j+1} \quad \text{for } i \le j. \tag{3.2}$$

Definition 3.3

A *semi-simplicial complex* K consists of a sequence $\{K_n | n \ge 0\}$ of disjoint sets together with set theoretic maps

$$d_i \colon K_{n+1} \to K_n \quad \text{for } 0 \le i \le n + 1,$$

$$s_i \colon K_n \to K_{n+1} \quad \text{for } 0 \le i \le n,$$

satisfying the following identities:

$$d_i d_j = d_{j-1} d_i \quad \text{for } i < j,$$

$$d_i s_j = s_{j-1} d_i \quad \text{for } i < j,$$

$$d_i s_j = \text{Id} \quad \text{for } i = j, \, j + 1,$$

$$d_i s_j = s_j d_{i-1} \quad \text{if } i > j + 1,$$

$$s_i s_j = s_{j+1} s_i \quad \text{if } i \le j. \tag{3.4}$$

Example 3.5

Let X be a topological space. For any $n \ge 0$ let $S_n(X)$ denote the set of singular n simplices of X. The maps $d_i \colon S_{n+1}(X) \to S_n(X)$, $s_i \colon S_n(X) \to S_{n+1}(X)$ are defined by

$$d_i(\sigma) = \sigma \circ \delta_i \quad \forall \sigma \in S_{n+1}(X),$$

$$s_i(f) = f \circ s_i \quad \forall f \in S_n(X).$$

Then $S(X) = \{S_n(X) | n \ge 0\}$ is a semi-simplicial complex called the *singular complex* of X.

Let K be any semi-simplicial complex. Let $\overline{K}$ be the topological union $\cup_{n \geq 0}(K_n \times \Delta_n)$ where each K_n is given the discrete topology. Consider the equivalence relation $\sim$ in $\overline{K}$ generated by

$$(\sigma_n, \partial_i t) \sim (d_i \sigma_n, t) \qquad \forall \, 0 \leq i \leq n, \, t \in \Delta_{n-1}, \, \sigma_n \in K_n,$$

$$(\sigma_n, s_i u) \sim (s_i \sigma_n, u) \qquad \forall \, 0 \leq i \leq n, \, u \in \Delta_{n+1}, \, \sigma_n \in K_n.$$

The identification space $|K| = \overline{K}/\sim$ is called the geometric realization of K. For any $x \in \Delta_n$ and $\sigma \in K_n$ the equivalence class of (σ, x) will be denoted by $|\sigma, x|$.

By a nondegenerate point of $\overline{K}$ we mean an element $(\sigma_n, x_n) \in \overline{K}$ with σ_n nondegenerate in K_n and $x_n \in \text{Int } \Delta_n$. It can be shown that every element of $\overline{K}$ is equivalent to a unique nondegenerate element. Using this fact the following result is proved in [12].

Theorem 3.6

$|K|$ is a CW-complex having one n cell corresponding to each nondegenerate n simplex of K.

Definition 3.7

Let $K = \{K_n | n \geq 0\}$, $L = \{L_n | n \geq 0\}$, be semi-simplicial complexes. By a semi-simplicial map $f: K \to L$ we mean a sequence of maps $f_n: K_n \to L_n$ $(n \geq 0)$ such that

$$
\begin{array}{ccc}
K_{n+1} & \xrightarrow{d_i} & K_n \\
{\scriptstyle f_{n+1}}\downarrow & & \downarrow{\scriptstyle f_n} \\
L_{n+1} & \xrightarrow{d_i} & L_n
\end{array}
\qquad \text{for } 0 \leq i \leq n+1,
$$

$$
\begin{array}{ccc}
K_n & \xrightarrow{s_i} & K_{n+1} \\
{\scriptstyle f_n}\downarrow & & \downarrow{\scriptstyle f_{n+1}} \\
L_n & \xrightarrow{s_i} & L_{n+1}
\end{array}
\qquad \text{for } 0 \leq i \leq n
$$

are all commutative.

Given a semi-simplicial map $f: K \to L$ there is an induced continuous map $|f|: |K| \to |L|$ given by $|f|(|\sigma, x|) = |f(\sigma), x|$ for each $\sigma \in K_n$, $x \in \Delta_n$. That $|f|$ is well defined is an immediate consequence of the commutativity of the preceding diagrams.

Let X be any topological space. Then there is a natural continuous map $\alpha_X: |S(X)| \to X$ given by $\alpha_X|\sigma, x| = \sigma(x) \; \forall \; \sigma \in S_n(X)$ and $x \in \Delta_n$. We now state without proof the following result.

Theorem 3.8

$\alpha_X: |S(X)| \to X$ is a weak homotopy equivalence.

For a proof refer to [12].

We are now ready to prove the result of Whitehead stated at the beginning of this section.

Theorem 3.9 (J. H. C. Whitehead)

Let X be a space dominated by a CW-complex. Then $\alpha_X: |S(X)| \to X$ is a homotopy equivalence. In particular X is homotopically equivalent to a CW-complex.

Proof. Let K be a CW-complex dominating X in homotopy. Let $f: X \to K$ and $g: K \to X$ be maps with $g \circ f \sim \mathrm{Id}_X$. Writing α for α_X, since α is a weak homotopy equivalence, from Corollary 2.15 we see that

$$\alpha_\#: \big[K, |S(X)|\big] \to [K, X] \tag{3.10}$$

and

$$\alpha_\#: \big[|S(X)|, |S(X)|\big] \to \big[|S(X)|, X\big] \tag{3.11}$$

are bijections. The bijective nature of (3.10) shows that there exists a map $\varphi: K \to |S(X)|$ such that $\sigma_\#[\varphi] = [g]$; equivalently $\alpha \circ \varphi \sim g$. Let $\beta = \varphi \circ f: X \to |S(X)|$. Then $\alpha \circ \beta = \alpha \circ \varphi \circ f \sim g \circ f \sim \mathrm{Id}_X$. Also

$$\alpha_\#[\beta \circ \alpha] = [\alpha \circ \beta \circ \alpha] = [\alpha \circ \varphi \circ f \circ \alpha] = [g \circ f \circ \alpha] = [\mathrm{Id}_X \circ \alpha]$$

$$= [\alpha] = \big[\alpha \circ \mathrm{Id}_{|S(X)|}\big] = \alpha_\#\big[\mathrm{Id}_{|S(X)|}\big].$$

From the bijective nature of (3.11) we get $\beta \circ \alpha \sim \mathrm{Id}_{|S(X)|}$. It follows that α is a homotopy equivalence with β as its homotopy inverse. $\square$

Corollary 3.12

Let X, Y be spaces dominated by CW-complexes. Then any weak homotopy equivalence $f: X \to Y$ is an actual homotopy equivalence.

Proof. Immediate consequence of Theorems 2.16 and 3.9. □

Corollary 3.13

Let X, Y be 1-connected spaces dominated by CW-complexes. Then $f: X \to Y$ is a homotopy equivalence if and only if it is a homology equivalence.

Proof. Immediate consequence of Theorem 1.14(ii) and Corollary 3.12. □

Lemma 3.14

Suppose (X, A, B) is a triple with (X, A) and (A, B) n-connected. Then (X, B) is n-connected.

Proof. Let $0 \le k \le n$ and $f: (E^k, S^{k-1}) \to (X, B)$ be any map. Since $B \subset A$ we may regard f as a map $(E^k, S^{k-1}) \to (X, A)$. Since (X, A) is n-connected, $f \sim \varphi$ rel. S^{k-1} where $\varphi: (E^k, S^{k-1}) \to (X, A)$ satisfies $\varphi(E^k) \subset A$. Since $f \sim \varphi$ rel. S^{k-1} we have $\varphi|S^{k-1} = f|S^{k-1}$. In particular $\varphi(S^{k-1}) \subset B$. Thus φ is a map $(E^k, S^{k-1}) \to (A, B)$. Since (A, B) is n-connected, it follows that $\varphi \sim h$ rel. S^{k-1} with $h(E^k) \subset B$. Then $f \sim h$ rel. S^{k-1} and $h(E^k) \subset B$. This shows that (X, B) is n-connected. □

By induction on r we immediately get the following:

Corollary 3.15

Let $(A_1, A_2, \ldots, A_r)$ be an r-ad $(r \ge 3)$ with (A_i, A_{i+1}) n-connected for $1 \le i \le r - 1$. Then (A_1, A_r) is n-connected.

Lemma 3.16

The pair (E^n, S^{n-1}) is $(n - 1)$-connected.

Proof. Let $f: (E^k, S^{k-1}) \to (E^n, S^{n-1})$ be any map with $0 \le k \le (n - 1)$. By the simplicial approximation theorem, there exist a simplicial subdivision (K, L) of (E^k, S^{k-1}) and a simplicial map $\varphi: (K, L) \to (E^n, S^{n-1}) = (\Delta_n, \dot{\Delta}_n)$ with $f \sim \varphi$ as maps of pairs. Since dim $K \le n - 1$, we get $\varphi(K) \subset \dot{\Delta}_n$.

However our aim is to get a map $\theta: (E^k, S^{k-1}) \to (E^n, S^{n-1})$ homotopic to f rel. S^{k-1} and satisfying $\theta(E^k) \subset S^{n-1}$. Let $F: (E^k \times I, S^{k-1} \times I) \to (E^n, S^{n-1})$ be a homotopy between f and φ, namely $F(x, 0) = f(x)$ and $F(x, 1) = \varphi(x) \ \forall \ x \in E^k$. Let $\Gamma: S^{k-1} \times I \cup E^k \times 1 \to S^{n-1}$ be given by

$\Gamma = F|S^{k-1} \times I \cup E^k \times 1$. Since $S^{k-1} \times I \cup E^k \times 1$ is a deformation retract of $E^k \times I$, there exists an extension $G: E^k \times I \to S^{n-1}$ of Γ. Let $M: E^k \times 1 \times I \cup S^{k-1} \times I \times I \cup E^k \times I \times \dot{I} \to E^n$ be given by

$$
\left.
\begin{aligned}
M(x, 1, t) &= F(x, t) \\
M(x, s, 0) &= f(x) \\
M(x, s, 1) &= G(x, s)
\end{aligned}
\right\} \qquad \forall\, x \in E^k,
$$

$$
M(a, s, t) = F(a, st) \qquad \forall\, a \in S^{k-1}.
$$

Clearly

$$
\begin{aligned}
M(x, 1, 0) &= F(x, 0) = f(x), \\
M(x, 1, 1) &= G(x, 1) = \varphi(x) = F(x, 1)
\end{aligned}
\qquad \text{for all } x \in E^k.
$$

Also

$$
\left.
\begin{aligned}
M(a, 1, t) &= F(a, t) \\
M(a, s, 0) &= F(a, 0) = f(a) \\
M(a, s, 1) &= F(a, 1) = G(a, 1)
\end{aligned}
\right\} \qquad \text{for all } a \in S^{k-1}.
$$

Thus M is well defined. Also $M(S^{k-1} \times I \times I \cup E^k \times I \times 1) \subset S^{n-1}$. Since $E^k \times 1 \times I \cup S^{k-1} \times I \times I \cup E^k \times I \times \dot{I}$ is a deformation retract of $E^k \times I \times I$ we get an extension $H: E^k \times I \times I \to E^n$ of M.

Consider $L: E^k \times I \to E^n$ given by $L(x, t) = H(x, 0, t)$. Then $L(x, 0) = H(x, 0, 0) = M(x, 0, 0) = f(x);$ $\quad L(x, 1) = H(x, 0, 1) = G(x, 0) \in S^{n-1}$. Moreover if $a \in S^{k-1}$ we get $L(a, t) = H(a, 0, t) = F(a, 0) = f(a)$. Thus L yields a homotopy $f \sim \theta$ rel. S^{k-1} where $\theta: (e^k, S^{k-1}) \to (e^n, S^{n-1})$ is given by $\theta(x) = L(x, 1)$. Cleary $\theta(E^k) \subset S^{n-1}$. $\quad\square$

Proposition 3.17

Let A be Hausdorff and $X = A \cup_{f_\alpha} \{\cup e_\alpha^n\}_{\alpha \in J}$. Then (X, A) is $(n - 1)$-connected.

Proof. Given a compact subset C of X it is easy to see that there exists a finite subset J' of J such that $C \subset A \cup_{f_\alpha} \{\cup e_\alpha^n\}_{\alpha \in J'}$. If this is not the case, the set $\{\alpha \in J | C \cap \mathrm{Int}\, e_\alpha^n \neq \varnothing\}$ will be infinite. If K is constituted by picking one element in each of $C \cap \mathrm{Int}\, e_\alpha^n$ for which $C \cap \mathrm{Int}\, e_\alpha^n \neq \varnothing$, then K is closed in C, hence compact, and also K is discrete—a contradiction since K is infinite.

Let $0 \leq k \leq n - 1$ and $f: (E^k, S^{k-1}) \to (X, A)$ be any map. Since $f(E^k)$ is compact, there exists a finite subset J' of J with $f(E^k) \subset A \cup_{f_\alpha} \{\cup e_\alpha^n\}_{\alpha \in J'}$. Writing X' for $A \cup_{f_\alpha} \{\cup e_\alpha^n\}_{\alpha \in J'}$ it suffices to show that $f: (E^k, S^{k-1}) \to (X', A)$ is "compressible," namely homotopic rel. S^{k-1} to a map $\theta: (E^k, S^{k-1}) \to (X', A)$ with $\theta(E^k) \subset A$. Thus we can assume $X = A \cup \cup_{i=1}^r e_i^n$. Writing B_j for $A \cup \cup_{i=1}^j e_i^n$ then $A \subset B_1 \subset \cdots \subset B_r = X$ and $B_{j+1} = B_j \cup e_{j+1}^n$ (also $B_1 = A \cup e_1^n$). From Corollary 3.15 we see that it suffices to prove that (B_{j+1}, B_j) is $(n-1)$-connected, where we set $B_0 = A$.

Thus we have only to show that (X, A) is $(n-1)$-connected when $X = A \cup e^n$. Let $h: (E^n, S^{n-1}) \to (e^n, \dot{e}^n)$ be the characteristic map. Let $u = h(0) \in \text{Int } e^n$, $U = \text{Int } e^n$, and $V = A \cup (e^n - \{u\})$. Then $\{U, V\}$ is an open covering of X and A is a strong deformation retract of V. Let $f: (E^k, S^{k-1}) \to (X, A)$ be any map with $0 \leq k \leq n - 1$. Then $\{F^{-1}(U), f^{-1}(V)\}$ is an open covering of E^k. If $\delta > 0$ is a Lebesgue number for this covering, there exists a simplicial structure (K, L) for (E^k, S^{k-1}) with mesh less than δ. Then every simplex of K is either in $f^{-1}(U)$ or in $f^{-1}(V)$. Let P be the union of simplices of K that lie in $f^{-1}(U)$ and Q be the union of simplices of K that lie in $f^{-1}(V)$. Then P, Q are subcomplexes of K and $L \subset Q$. Moreover $f|P: (P, P \cap Q) \to (U, U - \{u\}) = (\text{Int } e^n, \text{Int } e^n - \{u\})$. The pair $(\text{Int } e^n, \text{Int } e^n - \{u\})$ is homeomorphic to the pair $(\text{Int } E^n, \text{Int } E^n - \{0\})$. If $E^n_{1/2}$ is the closed disk of radius $1/2$, then $(\text{Int } E^n, \text{Int } E^n - \{0\})$ is homotopically equivalent to $(E^n_{1/2}, E^n_{1/2} - \{0\})$ which in turn is homotopically equivalent to $(E^n_{1/2}, \dot{E}^n_{1/2})$, hence $(n-1)$-connected by Lemma 3.16. Clearly $\dim(P, P \cap Q) \leq n - 1$. Any simplicial pair being trivially a CW-pair from Proposition 2.11 we see that $f: (P, P \cap Q) \to (U, U - \{u\})$ is compressible. Since $f(Q) \subset V$, $P \cup Q = K$, and $U \cup V = X$ we see that $f: (K, Q) \to (X, V)$ is compressible. From $L \subset Q$, we immediately see that $f: (K, L) \to (X, V)$ is compressible. Since A is a deformation retract of V we see that $f: (K, L) \to (X, A)$ is compressible. $\square$

Theorem 3.18

Let (X, A) be a CW-pair. For each $q \geq 0$ let $X^q = A \cup$ cells of X of dim $n \leq q$. Then (X^q, X^n) is n-connected for all $q \geq n$. Also (X, X^n) is n-connected.

Proof. The pair (X^q, X^{q-1}) is $(q-1)$-connected by Proposition 3.17, hence n-connected if $q > n$. Thus (X^{n+1}, X^n) and $(X^{n+2}, X^{n+1}), \ldots, (X^q, X^{q-1})$ are all n-connected if $q > n$. From Corollary 3.15 we see that (X^q, X^n) is n-connected if $q > n$. As for (X^n, X^n) it is clearly ∞-connected.

We know that any compact subset of X lies in a certain X^q. Hence if $f: (E^k, S^{k-1}) \to (X, X^n)$ is a given map, we get $f(E^k) \subset X^q$ for some q (we can assume $q > n$). Since (X^q, X^n) is n-connected, we see that whenever $0 \le k \le n$, f can be compressed. This proves that (X, X^n) is n-connected. $\square$

Definition 3.19

Let X, Y be CW-complexes. A map $f: X \to Y$ is said to be *cellular* if $f(X_n) \subset Y_n$ for all $n \ge 0$.

A homotopy $F: X \times I \to Y$ is said to be cellular if $F((X \times I)_n) \subset Y_n$ for all $n \ge 0$. [Equivalently $F(X_n \times 0 \cup X_n \times 1 \cup X_{n-1} \times I) \subset Y_n$ for all $n \ge 0$.]

Theorem 3.20 (Cellular Approximation Theorem)

Let X, Y be CW-complexes and $f: X \to Y$ a continuous map. Let A be a subcomplex of X. Suppose $f|A$ is cellular. Then f is homotopic relative to A to a cellular map $g: X \to Y$.

Proof. By induction on n we define a sequence of homotopies $F_n: X_n \times I \to Y$ satisfying the following conditions:

(i) $F_n(x, 0) = f(x)$ for all $x \in X_n$.
(ii) $F_n(a, t) = f(a)$ for all $a \in A_n = A \cap X_n$.
(iii) $F_n|X_{n-1} \times I = F_{n-1}$ for $n \ge 1$.
(iv) $F_n(X_n \times 1) \subset Y_n$.

We start the induction at $n = 0$. For each 0 cell σ of X not in A let $g_0(\sigma)$ equal a vertex in the arc component of Y containing $f(\sigma)$. Let α_σ be a path in Y joining $f(\sigma)$ to $g_0(\sigma)$.

Define $F_0: X_0 \times I \to Y$ by

$$F_0(a, t) = f(a) \qquad \forall\, a \in A_0 = A \cap X_0,$$

$$F_0(x, t) = \alpha_x(t) \quad \text{for any } x \in X_0 - A_0.$$

Clearly $f_0(X_0 \times 1) \subset Y_0$.

Let $n \ge 0$ and assume $F_0, \ldots, F_n$ constructed satisfying the preceding conditions (i)–(iv).

Let e_α^{n+1} be an $(n + 1)$ cell of X not in A and

$$g_\alpha: (E^{n+1}, S^n) \to (e_\alpha^{n+1}, \dot{e}_\alpha^{n+1})$$

be the characteristic map of e_α^{n+1}. Consider $\Psi_\alpha: E^{n+1} \times 0 \cup S^n \times I \to Y$

given by

$$\Psi_\alpha(u,0) = f(g_\alpha(u)) \qquad \forall \, u \in E^{n+1},$$

$$\Psi_\alpha(u,t) = F_n(g_\alpha(u),t) \qquad \forall \, u \in S^n.$$

Then $\Psi_\alpha(S^n \times 1) = F_n(g_\alpha(S^n) \times 1) \subset Y_n$ by (iv). In particular $\Psi_\alpha(S^n \times 1) \subset Y_{n+1}$. We know that (Y, Y_{n+1}) is $(n+1)$-connected (Theorem 3.18). Hence there exists an extension $\Phi_\alpha \colon E^{n+1} \times I \to Y$ of Ψ_α satisfying $\Phi_\alpha(E^{n+1} \times 1) \subset Y_{n+1}$. It is clear that Φ_α induces a map $G_\alpha \colon e_\alpha^{n+1} \times I \to Y$ satisfying $G_\alpha \circ (g_\alpha \times \mathrm{Id}_I) = \Phi_\alpha$. Then $G_\alpha | \dot{e}_\alpha^{n+1} \times I = F_n | \dot{e}_\alpha^{n+1} \times I$ and $G_\alpha(e_\alpha^{n+1} \times 1) \subset Y_{n+1}$. Then $F_{n+1} \colon X_{n+1} \times I \to Y$ defined by

$$F_{n+1}(a,t) = f(a) \qquad \forall \, a \in A_{n+1} = A \cap X_{n+1},$$

$$F_{n+1} | X_n \times I = F_n,$$

and

$$F_{n+1} | e_\alpha^{n+1} \times I = G_\alpha$$

for every $(n+1)$ cell of X not in A satisfies (i)–(iv) with n replaced by $(n+1)$.

Then $F \colon X \times I \to Y$ given by $F | X_n \times I = F_n$ is a homotopy of f relative to A and $g \colon X \to Y$ given by $g(x) = F(x,1) \; \forall \, x \in X$ is cellular. $\square$

Corollary 3.21

Let X, Y be CW-complexes, A a subcomplex of X, and $f, g \colon X \to Y$ cellular maps. Suppose there exists a homotopy $G \colon X \times I \to Y$ between f and g further satisfying the condition that $G | A \times I$ is cellular. Then there exists a cellular homotopy $F \colon X \times I \to Y$ between f and g with $F | A \times I = G$.

Proof. $G \colon X \times I \to Y$ satisfies the condition that $G | X \times 0 \cup X \times 1 \cup A \times I$ is cellular. By Theorem 3.20, there exists a map $\Gamma \colon X \times I \times I \to Y$ satisfying

$$\Gamma(x,t,0) = G(x,t) \qquad \forall \, x \in X, t \in I,$$

$$\left. \begin{array}{l} \Gamma(x,0,s) = f(x) \\[4pt] \Gamma(x,1,s) = g(x) \end{array} \right\} \qquad \forall \, x \in X, s \in I,$$

$$\Gamma(a,t,s) = G(a,t) \qquad \forall \, a \in A, t \in I, s \in I,$$

and the map $F: X \times I \to Y$ given by

$$F(x, t) = \Gamma(x, t, 1) \quad \text{cellular.}$$

F clearly satisfies the requirements of Corollary 3.21. $\square$

4   AN EXAMPLE

In this section we will give an example of a homology equivalence $f: X \to Y$ of 0-connected spaces with the additional property that $f_*: \pi_1(X) \simeq \pi_1(Y)$ which is not a weak equivalence.

Let k be any integer greater than 1 and $A = S^3 \cup_k e^4$ where e^4 is attached to S^3 by means of a map $\varphi: S^3 \to S^3$ of degree k. Then A is a Moore space of type $(Z/kZ, 3)$. In fact A is a finite simply connected CW-complex satisfying $H_3(A, Z) \simeq Z/kZ$ and $H_j(A, Z) = 0$ for $j \geq 1$ and $j \neq 3$. Let Q denote the additive group of rational numbers and

$$X = K(Q, 1) \vee A,$$

$$Y = K(Q, 1) \times A.$$

We can actually choose $K(Q, 1)$ to be a countable CW-complex so that X and Y themselves are CW-complexes. Let $f: K(Q, 1) \vee A \to K(Q, 1) \times A$ denote the inclusion map. It is clear that $f_*: \pi_1(X) = Q \to \pi_1(Y) = Q$ is the identity map. Observe that $H_i(K(Q, 1), Z) = 0$ for $i \geq 2$. From this we see that $H_1(X) \simeq Q$, $H_3(X) \simeq Z/kZ$, and $H_j(X) = 0$ for all *other* $j \geq 1$. From the Kunneth formulas we have $H_1(Y) \simeq Q$, $H_2(Y) = 0$, $H_3(Y) \simeq Z/kZ$, $H_4(Y) \simeq Q \otimes_Z (Z/kZ) = 0$, and $H_j(Y) = 0$ for $j \geq 5$.

It is now straightforward to see that $f: X \to Y$ is a homology equivalence further satisfying the condition that $f_*: \pi_1(X) \simeq \pi_1(Y)$.

We claim that f is not a weak equivalence because if f were a weak equivalence, then by Theorem 2.16 it would be a homotopy equivalence. If this were the case, the universal coverings $\tilde{X}, \tilde{Y}$ of X and Y would be homotopy equivalent.

Clearly, $\tilde{X}$ is homotopy equivalent to $\bigvee_{r \in Q} M(Z/kZ, 3)$, a wedge of Moore spaces of type $(Z/kZ, 3)$ indexed by Q whereas $\tilde{Y}$ is homotopy equivalent to $M(Z/kZ, 3)$. Thus $\tilde{X}$ is not homotopically equivalent to $\tilde{Y}$.

5   POSTNIKOV SYSTEMS

We will explain in this section the method of obtaining a *Postnikov system* for any given arcwise-connected space X.

Let X be a 0-connected space and n an integer greater than or equal to 1.

Lemma 5.1

There exists a space $X^{[n]}$ and an inclusion $i_n: X \to X^{[n]}$ such that

(i) $X^{[n]}$ is obtained from X by attaching cells of dimension greater than or equal to $n + 2$.

(ii) $\pi_j(X^{[n]}) = 0$ for $j > n$.

(iii) $\pi_j(X) \xrightarrow[\approx]{(i_n)_*} \pi_j(X^{[n]})$ for $j \leq n$.

Proof. Let $(a_\alpha)_{\alpha \in J_{n+1}}$ generate $\pi_{n+1}(X)$ and let $f_\alpha: S^{n+1} \to X$ represent a_α. Let $X' = X \cup_{f_\alpha} \{e_\alpha^{n+2}\}_{\alpha \in J_{n+1}}$. Then in the exact homotopy sequence of the pair (X', X),

$$\pi_{n+2}(X', X) \xrightarrow{\partial} \pi_{n+1}(X) \longrightarrow \pi_{n+1}(X') \longrightarrow \pi_{n+1}(X', X)$$
$$\xrightarrow{\partial} \pi_n(X) \longrightarrow \pi_n(X') \longrightarrow \cdots$$

we have $\pi_i(X', X) = 0$ for $i \leq n + 1$.

$\pi_{n+2}(X', X)$ is generated by w_α over $\pi_1(X)$ where $w_\alpha = [g_\alpha]$ where $g_\alpha: (D^{n+1}, S^{n+1}) \to (X', X)$ is the obvious extension of f_α, namely f_α is extended to a map of D^{n+1} into $\bar{e}_\alpha^{n+2}$. Then $\partial w_\alpha = [f_\alpha] \in \pi_{n+1}(X)$. Since $\{a_\alpha\}$ generate $\pi_{n+1}(X)$ we see from the preceding exact sequence

$$\pi_{n+1}(X') = 0, \qquad \pi_j(X) \simeq \pi_j(X') \quad \text{for } j \leq n$$

under the map induced by the inclusion $X \to X'$.

X' is obtained from X by attaching $(n + 2)$ cells. The homotopy groups π_j for $j \leq n$ are unaltered and π_{n+1} is completely killed. Repeating this construction we get a space

$$X^{[n]} = X \cup \{e_\alpha^{n+2}\}_{\alpha \in J_{n+1}} \cup \{e_\beta^{n+3}\}_{\beta \in J_{n+2}} \cup \cdots$$

with $\pi_i(X^{[n]}) = 0$ for $i \geq n + 1$,

$$\pi_i(X) \xrightarrow{i_{n*}} \pi_i(X^{[n]}) \quad \text{isomorphism for } i \leq n$$

where $i_n: X \to X^{[n]}$.

Proposition 5.2

Let $1 \leq m \leq n$, $f: X \to Y$ a map of 0-connected spaces,

$$i_n: X \to X^{[n]},$$

$$i_m: Y \to Y^{[m]},$$

satisfy the conditions of Lemma 5.1. Then there exists a map $f_{n,m}: X^{[n]} \to Y^{[m]}$ *such that*

$$
\begin{array}{ccc}
X & \xrightarrow{\ f\ } & Y \\
{\scriptstyle i_n}\downarrow & & \downarrow{\scriptstyle i_m} \\
X^{[n]} & \xrightarrow[\ f_{n,m}\ ]{} & Y^{[m]}
\end{array}
$$

is commutative. Moreover any two such $f_{n,m}$ *are homotopic relative to* X.

Proof. The existence of such an $f_{n,m}$ is an immediate consequence of the facts that $X^{[n]}$ is obtained from X by attaching cells of dimension greater than or equal to $n+2$ and that $\pi_j(Y^{[m]}) = 0$ for $j \geq n+1$ (in fact for $j \geq m+1$ as well).

Let

$$f_{n,m}: X^{[n]} \to Y^{[m]},$$

$$h_{n,m}: X^{[n]} \to Y^{[m]},$$

be two such maps.

Let $\varphi: X^{[n]} \times \dot{I} \cup X \times I \to Y^{[m]}$ be defined by

$$\varphi(x, t) = f(x) \qquad \forall\, x \in X, t \in I,$$

$$\left.\begin{array}{l} \varphi(u, 0) = f_{n,m}(u) \\[4pt] \varphi(u, 1) = h_{n,m}(u) \end{array}\right\} \qquad \forall\, u \in X^{[n]}.$$

$(X^{[n]} \times I,\ X^{[n]} \times \dot{I} \cup X \times I)$ is a relative CW-complex with cells in dimensions greater than or equal to $n+3$. Since $\pi_j(Y^{[m]}) = 0$ for $j \geq m+1$ (and $m \leq n$) we see that φ can be extended to a map $F: X^{[n]} \times I \to Y^{[m]}$. Any such F is a homotopy between $f_{n,m}$ and $h_{n,m}$ relative to X. This completes the proof of Proposition 5.2. $\square$

Let $X \xrightarrow{\ i_n\ } X^{[n]}$ and $X \xrightarrow{\ \mu_n\ } \overline{X}^{[n]}$ be two inclusions satisfying conditions (i)–(iii) of Lemma 5.1. Then from Proposition 5.2 we see that $\exists$ maps

$$f_{n,n}: X^{[n]} \to \overline{X}^{[n]},$$

$$g_{n,n}: \overline{X}^{[n]} \to X^{[n]},$$

such that

$$
\begin{array}{ccc}
X & \xrightarrow{\ i_n\ } & X^{[n]} \\
\| & & \downarrow{\scriptstyle f_{n,\,n}} \\
X & \xrightarrow{\ \mu_n\ } & \overline{X}^{[n]} \\
\| & & \downarrow{\scriptstyle g_{n,\,n}} \\
X & \longrightarrow & X^{[n]}
\end{array}
$$

is a commutative diagram. But $\mathrm{Id}_{X^{[n]}}\colon X^{[n]} \to X^{[n]}$ and $\mathrm{Id}_{\overline{X}^{[n]}}\colon \overline{X}^{[n]} \to \overline{X}^{[n]}$ clearly make

$$
\begin{array}{ccc}
X & \xrightarrow{\ i_n\ } & X^{[n]} \\
\| & & \downarrow{\scriptstyle \mathrm{Id}_{X^{[n]}}} \\
X & \xrightarrow{\ i_n\ } & X^{[n]}
\end{array}
\quad,\quad
\begin{array}{ccc}
X & \xrightarrow{\ \mu_n\ } & \overline{X}^{[n]} \\
\| & & \downarrow{\scriptstyle \mathrm{Id}_{\overline{X}^{[n]}}} \\
X & \xrightarrow{\ \mu_n\ } & \overline{X}^{[n]}
\end{array}
$$

commutative. Proposition 5.2 now implies

$$
g_{n,\,n} \circ f_{n,\,n} \sim \mathrm{Id}_{X^{[n]}} \text{ rel. } X,
$$

$$
f_{n,\,n} \circ g_{n,\,n} \sim \mathrm{Id}_{\overline{X}^{[n]}} \text{ rel. } X.
$$

Thus the homotopy type of $X^{[n]}$ (rel. X) is completely determined.

When $f\colon X \to Y$ is a map of 0-connected spaces, any map $f_{n,\,n}$ making

$$
\begin{array}{ccc}
X & \xrightarrow{\ i_n\ } & X^{[n]} \\
{\scriptstyle f}\downarrow & & \downarrow{\scriptstyle f_{n,\,n}} \\
Y & \xrightarrow{\ i_n\ } & Y^{[n]}
\end{array}
$$

commutative will be denoted by $f^{[n]}\colon X^{[n]} \to Y^{[n]}$. As seen already $f^{[n]}$ is unique up to homotopy rel. X. Using Proposition 5.2 we see that $\exists$ maps $\varphi_n\colon X^{[n]} \to X^{[n-1]}$ for all $n \geq 2$ such that

$$
\begin{array}{ccc}
X & \xrightarrow{\ i_n\ } & X^{[n]} \\
\| & & \downarrow{\scriptstyle \varphi_n} \\
X & \xrightarrow{\ i_{n-1}\ } & X^{[n-1]}
\end{array}
$$

is commutative. Thus we get a "tower"

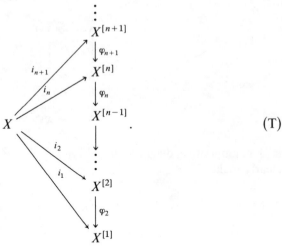

$$\tag{T}$$

Clearly $X^{[1]}$ is a $K(\pi_1(X), 1)$ space.

Using the standard procedure we replace each of the maps $\varphi_2, \varphi_3, \varphi_4, \ldots$ by a (Hurewicz) fibration without altering the homotopy types of spaces involved.

Then we get a diagram

$$
\begin{array}{ccc}
\vdots & & \vdots \\
\downarrow & & \downarrow \\
X^{[n+1]} & \xrightarrow{\alpha_{n+1}} & P_{n+1}X \\
\varphi_{n+1}\downarrow & & \downarrow\rho_{n+1} \\
X^{[n]} & \xrightarrow{\alpha_n} & P_nX \\
\varphi_n\downarrow & & \downarrow\rho_n \\
X^{[n-1]} & \longrightarrow & P_{n-1}X \\
\vdots & & \vdots \\
\downarrow & & \downarrow \\
X^{[2]} & \xrightarrow{\alpha_2} & P_2X \\
\varphi_2\downarrow & & \downarrow\rho_2 \\
X^{[1]} & \xrightarrow{\alpha_1=\mathrm{Id}} & P_1X = X^{[1]}
\end{array}
$$

which is homotopy commutative, with the additional properties that each ρ_n is a Hurewicz fibration.

The commutativity of the tower (T) implies that

$$\varphi_{n*}: \pi_i(X^{[n]}) \simeq \pi_i(X^{[n-1]}) \quad \text{for } i \leq n - 1.$$

Hence, for any $n \geq 2$,

$$\rho_{n*}: \pi_i(P_n X) \simeq \pi_i(P_{n-1} X) \quad \text{for } i \leq n - 1.$$

It follows from the homotopy exact sequence of the fibration ρ_n that the fiber of ρ_n is a $K(\pi_n(X), n)$ space. If $h_n = \alpha_n \circ i_n: X \to P_n X$ then for $n \geq 2$,

$$
\begin{array}{ccc}
X & \xrightarrow{h_n} & P_n X \\
\| & & \downarrow{\scriptstyle \rho_n} \\
X & \xrightarrow{h_{n-1}} & P_{n-1} X
\end{array}
$$

is commutative up to homotopy. Using the fact that ρ_n is a fibration, we get successively maps $\mu_1: X \to P_1 X$, $\mu_2: X \to P_2 X, \ldots, \mu_n: X \to P_n X, \ldots$ such that

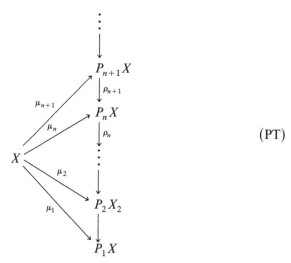

(PT)

is actually commutative:

$$\mu_{n*}: \pi_i(X) \simeq \pi_i(P_n X) \quad \text{for } i \leq n,$$

$$\pi_k(P_n X) = 0 \quad \text{for } k \geq n + 1.$$

The "tower" (PT) is called the Postnikov tower of X. Note that μ_n's need not be inclusions.

Chapter Two

E. Dror's Generalization of Whitehead's Theorem

In this chapter we deal with E. Dror's generalization of Whitehead's theorem [6]. $\underline{N}$ will denote the category of nilpotent groups and all group homomorphisms between them. We will follow the usual convention of writing $f: G \rightarrowtail H$ if it is a monomorphism of groups and $f: G \twoheadrightarrow H$ if it is an epimorphism of groups.

1 STALLINGS' THEOREM [20]

Proposition 1.1

Let $G \xrightarrow{f} K$ be an epimorphism of groups with $\ker f = N$. Then $\exists$ a "canonical" exact sequence

$$H_2(G) \xrightarrow{f_*} H_2(K) \longrightarrow N/[G, N] \longrightarrow H_1(G) \xrightarrow{f_*} H_1(K) \longrightarrow 0.$$

The word "canonical" means the following: If

$$
\begin{array}{ccc}
G & \xrightarrow{\ f\ } & K \\
{\scriptstyle \varphi}\downarrow & & \downarrow{\scriptstyle \psi} \\
G' & \xrightarrow[\ f'\ ]{} & K'
\end{array}
$$

is a commutative diagram in the category of groups with $\ker f = N$ *and* $\ker f' = N'$, *then*

$$H_2(G) \xrightarrow{f_*} H_2(K) \longrightarrow N/[G, N] \longrightarrow H_1(G) \xrightarrow{f_*} H_1(K) \longrightarrow 0$$

$$\varphi_* \downarrow \qquad \psi_* \downarrow \qquad \bar{\varphi} \downarrow \qquad \varphi_* \downarrow \qquad \psi_* \downarrow$$

$$H_2(G') \xrightarrow{f_*'} H_2(K') \longrightarrow N'/[G', N'] \longrightarrow H_1(G') \xrightarrow{f_*'} H_1(K') \longrightarrow 0$$

is commutative. Here $\bar{\varphi}$ *is induced by* φ.

Proof. We have the Lyndon–Hochschild–Serre spectral sequence with

$$E^2_{p,q} = H_p\big(G/N; H_q(N)\big),$$

$E^\infty_{p,q}$ = associated graded corresponding to a nice filtration on $H_*(G)$.

We make G act on N on the right via

$$x * g = g^{-1}xg \qquad \forall\, x \in N, g \in G.$$

This gives a right G action on $H_*(N)$. The action of N on itself is by inner automorphisms and the induced action of N on $H_*(N)$ is known to be trivial. Hence one gets an action of G/N on $H_*(N)$. In $E^2_{p,q} = H_p(G/N; H_q(N))$ it is this action of G/N on $H_*(N)$ that is to be considered. From $E^2_{4,-1} = 0$ and the definition of E^3 we see that $E^3_{2,0} = \ker d^2 : E^2_{2,0} \to E^2_{0,1}$. Also from

$$0 = E^r_{2+r,1-r} \quad \text{and} \quad E^r_{2-r,r-1} = 0 \quad \text{for } r \geq 3,$$

we get $E^3_{2,0} = E^\infty_{2,0}$. Hence $E^\infty_{2,0}$ can be identified with $\ker d^2 : E^2_{2,0} \to E^2_{0,1}$. However $E^\infty_{2,0} = F_{2,0}/F_{1,1} = [H_2(G)]/[F_1 H_2(G)]$. Thus there exists an epimorphism $H_2(G) \twoheadrightarrow E^\infty_{2,0}$. Thus one gets the exact sequence $H_2(G) \to E^2_{2,0} \xrightarrow{d^2} E^2_{0,1} \to E^2_{0,1}/\mathrm{Im}\, d^2 \to 0$ where $H_2(G) \to E^{2,0}$ is the composite of $H_2(G) \twoheadrightarrow E^\infty_{2,0}$ with the inclusion $E^\infty_{2,0} \to E^2_{2,0}$ (observing that $E^\infty_{2,0} = \ker d^2$). From $E^2_{-2,2} = 0$ we see that $E^3_{0,1} = E^2_{0,1}/\mathrm{Im}\, d^2$. Also $E^r_{r,2-r}$ and $E^r_{-r,r}$ are 0 for $r \geq 3$. It follows that $E^3_{0,1} = E^\infty_{0,1}$. Thus we can identify $E^2_{0,1}/\mathrm{Im}\, d^2$ with $E^\infty_{0,1}$. But $E^\infty_{0,1} = [F_0 H_1(G)]/[F_{-1} H_1(G)] = F_0 H_1(G)$. Thus $E^\infty_{0,1} \subset H_1(G)$. Moreover $[H_1(G)]/E^\infty_{0,1} = [H_1(G)]/[F_0, H_1(G)] = [F_1 H_1(G)]/[F_0 H_1(G)] = E^\infty_{1,0}$.

Hence we get the exact sequence

$$H_2(G) \longrightarrow E^2_{2,0} \xrightarrow{d^2} E^2_{0,1} \longrightarrow H_1(G) \longrightarrow E^\infty_{1,0} \longrightarrow 0. \qquad (*)$$

Now

$$E^2_{2,0} = H_2\big(G/N, H_0(N)\big) = H_2\big(G/N\big). \qquad (a)$$

The action of G/N on $H_0(N) = Z$ is trivial; also

$$E_{0,1}^2 = H_0(G/N, H_1(N))$$

$$= H_0(G/N; N/[N, N])$$

$$= Z \bigotimes_{G/N} \frac{N}{[N, N]} = \frac{N}{[G, N]}. \tag{b}$$

$E_{1+r,1-r}^r$ and $E_{1-r,r-1}^r$ are 0 for $r \geq 2$. It follows that $E_{1,0}^2 = E_{1,0}^\infty$. Thus

$$E_{1,0}^\infty = H_1(G/N, H_0(N)) = H_1(G/N). \tag{c}$$

Substituting in (a)–(c) the corresponding groups, the exact sequence $(*)$ reduces to the exact sequence

$$H_2(G) \to H_2(G/N) \to N/[G, N] \to H_1(G) \to H_1(G/N) \to 0.$$

The naturality is immediate from the naturality of the Hochschild–Serre spectral sequence. $\square$

Theorem 1.2

Let $f: G \to K$ be a homomorphism of groups with

$$f_*: H_1(G) \simeq H_1(K),$$

$$f_*: H_2(G) \twoheadrightarrow H_2(K).$$

Then the map $\bar{f}: G/\Gamma^i(G) \to K/\Gamma^i(K)$ induced by f is an isomorphism for every integer $i \geq 1$. Here $\Gamma^i(G)$ denotes the lower central series of G, namely $\Gamma^1(G) = G$ and $\Gamma^{i+1}(G) = [G, \Gamma^i(G)]$ for $i \geq 1$.

Proof. For $i = 1$ the result is clear there. For $i = 2$, $G/\Gamma^2(G) = H_1(G)$, $K/\Gamma^2(K) = H_1(K)$, and the result is true because of the first assumption. Let $i > 2$ and assume inductively that the result is true for $i - 1$ in place of i. From the epimorphisms

$$G \twoheadrightarrow G/\Gamma^{i-1}(G),$$

$$K \twoheadrightarrow K/\Gamma^{i-1}(K),$$

we get the following commutative diagram with exact rows:

$$H_2(G) \longrightarrow H_2(G/\Gamma^{i-1}(G)) \longrightarrow \Gamma^{i-1}(G)/\Gamma^i(G) \longrightarrow H_1(G) \longrightarrow H_1(G/\Gamma^{i-1}(G)) \longrightarrow 0$$

$$H_2(K) \longrightarrow H_2(K/\Gamma^{i-1}(K)) \longrightarrow \Gamma^{i-1}(K)/\Gamma^i(K) \longrightarrow H_1(K) \longrightarrow H_1(K/\Gamma^{i-1}(K)) \longrightarrow 0$$

From the five lemma we see that $\Gamma^{i-1}(G)/\Gamma^i(G) \simeq \Gamma^{i-1}(K)/\Gamma^i(K)$.
Again from

$$1 \longrightarrow \Gamma^{i-1}(G)/\Gamma^i(G) \longrightarrow G/\Gamma^i(G) \longrightarrow G/\Gamma^{i-1}(G) \longrightarrow 1$$

$$1 \longrightarrow \Gamma^{i-1}(K)/\Gamma^i(K) \longrightarrow K/\Gamma^i(K) \longrightarrow K/\Gamma^{i-1}(K) \longrightarrow 1$$

we get an isomorphism $G/\Gamma^i(G) \simeq K/\Gamma^i(K)$. □

Corollary 1.3

Let $f: G \to K$ be a map in $\underline{\underline{N}}$ with

$$H_1(f): H_1(G) \simeq H_1(K),$$

$$H_2(f): H_2(G) \twoheadrightarrow H_2(K).$$

Then $f: G \simeq K$.

Proof. $\exists$ an i with $\Gamma^i(G) = \{1\} = \Gamma^i(K)$. Hence from Theorem 1.2 we get $G \overset{f}{\simeq} K$. □

2 NILPOTENT ACTION OF A GROUP ON ANOTHER GROUP

We now want to introduce the notion of nilpotent action of a group π on another group G. Let π act on G through $\theta: \pi \to \text{Aut } G$.

Definition 2.1

The *lower π-central series* of the π group G is defined as follows: $\Gamma^1_\pi(G) = G$, $\Gamma^2_\pi(G)$ is the smallest normal π subgroup of G containing $[G, G]$, and $\{\theta(x)(g) \cdot g^{-1} | x \in \pi, \ g \in G\}$. For $k \geq 3$, $\Gamma^k_\pi(G)$ is the smallest normal π subgroup of G containing $[G, \Gamma^{k-1}_\pi(G)]$ and $\{\theta(x)(g) \cdot g^{-1} | x \in \pi, \ g \in \Gamma^{k-1}_\pi(G)\}$.

Definition 2.2

The action of π on G is said to be nilpotent if $\exists$ an integer k with $\Gamma_\pi^k(G) = \{1\}$.

Observe that $\Gamma_\pi^k(G) \supseteq \Gamma^k(G)$. Thus if the action of π on G is nilpotent, automatically the group G has to be nilpotent. In other words if G is a nonnilpotent group and if π acts trivially on G, this action of π on G is not a nilpotent action.

Suppose M is a π-module, namely M is an abelian group on which π acts. Then $\Gamma_\pi^1(M) = M$ and $\Gamma_\pi^{k+1}(M) = (I\pi)^k M$ for any integer $k \geq 1$, where $I\pi$ is the augmentation ideal in $Z\pi$. Hence a π-module M is a nilpotent π-module (or the action of π on M is nilpotent) $\Leftrightarrow (I\pi)^k M = 0$.

In case $G = \pi$ and π acts on itself by inner automorphisms $\Gamma_\pi^k(\pi) = \Gamma^k(\pi)$ for every integer $k \geq 1$. Hence π acts nilpotently on π under inner automorphism action $\Leftrightarrow \pi$ is a nilpotent group.

Lemma 2.3

Let M be a π-module. Then the action of π on M is nilpotent $\Leftrightarrow \exists$ a finite filtration $M = M_0 \supset M_1 \supset M_2 \supset \cdots \supset M_k = 0$ of M by π-submodules such that M_i/M_{i+1} is a trivial π-module $(0 \leq i \leq k - 1)$.

Proof. Assume $(I\pi)^k M = 0$. Then if we set $M_i = (I\pi)^i M$ for $i \geq 1$, $M = M_0 \supset M_1 \supset \cdots \supset M_k = 0$ is a filtration with M_i/M_{i+1} a trivial π-module.

Also if N is π-submodule of M with M/N a trivial π-module, then $(I\pi)M \subset N$. Hence if $M = M_0 \supset M_1 \supset \cdots \supset M_k = 0$ is a filtration of M with each M_i/M_{i+1} a trivial π-module $(0 \leq i \supset k - 1)$ we have $(I\pi)M_i \subset M_{i+1}$ for every i and hence $(I\pi)^k M = (I\pi)^k M_0 \subset M_k = 0$. $\square$

Definition 2.4

For any π group G the π completion $\hat{G}$ is defined to be the inverse limit of

$$\cdots \to G/\Gamma_\pi^k(G) \to \cdots G/\Gamma_\pi^3(G) \to G/\Gamma_\pi^2(G) \to G/\Gamma_\pi^1(G) = \{1\}.$$

There is an obvious homomorphism $i: G \to \hat{G}$ where $i(g) = ((\text{the coset of } g$ in $G/\Gamma_\pi^k(G))_k) \in \hat{G}$. The kernel of $i = \bigcap_{k \geq 1} \Gamma_\pi^k(G)$. This will be denoted by $\Gamma_\pi^\infty(G)$ or when the group π and its action on G are clear from the context we simply write $\Gamma_\infty(G)$.

Let π act on G, π' act on G', and $\alpha: \pi \to \pi'$ be a homomorphism. A homomorphism $f: G \to G'$ is called α compatible or simply an α homomorphism if for any $x \in \pi$ and $g \in G$,

$$f(x \cdot g) = \alpha(x) \cdot f(g)$$

where $x \cdot g$ denotes the element of G obtained from g by the action of x on g. Similarly $\alpha(x) \cdot f(g)$ is the element of G' obtained from $f(g)$ by the action of $\alpha(x)$. If $\alpha = \mathrm{Id}_\pi$, $f: G \to G'$ is the usual notion of π homomorphism.

If $f: G \to G'$ is an α homomorphism, it is clear that $f(\Gamma_\pi^k(G)) \subset \Gamma_{\pi'}^k(G')$ for every $k \geq 1$, hence $\exists$ induced maps $\bar{f}_k: G/\Gamma_\pi^k(G) \to G'/\Gamma_{\pi'}^k(G')$. These $\bar{f}_k$'s induce a map $\hat{f}: \hat{G} \to \hat{G}'$. Quite often from the context α will be clear.

For any ordinal β, we define $\Gamma_\pi^\beta(G)$ inductively as follows. If β is a nonlimiting ordinal, then $\Gamma_\pi^\beta(G)$ equals the smallest normal π subgroup of G containing $[G, \Gamma_\pi^{\beta-1}(G)]$ and $\{\theta(x)(g) \cdot g^{-1} | x \in \pi, g \in \Gamma_\pi^{\beta-1}(G)\}$ where $\theta: \pi \to \mathrm{Aut}\, G$ defines the action of π on G. If β is a limiting ordinal, $\Gamma_\pi^\beta(G) = \bigcap_{\gamma < \beta} \Gamma_\pi^\gamma(G)$.

Definition 2.5

Let G be a π group. We will call G π-perfect if $\Gamma_\pi^2(G) = G$.

Let G be any π group and $\{G_\alpha\}_{\alpha \in J}$ any family of π-perfect π subgroups of G. Let K be the π subgroup of G generated by $\bigcup_{\alpha \in J} G_\alpha$. Then K is a π-perfect subgroup of G. In fact $K \supset G_\alpha \Rightarrow \Gamma_\pi^2(K) \supset \Gamma_\pi^2(G_\alpha) = G_\alpha$. Hence $\Gamma_\pi^2(K) \supset \cup G_\alpha$ and hence $\Gamma_\pi^2(K) \supset K$ or $K = \Gamma_\pi^2(K)$. Because of this there is a unique largest π-perfect subgroup of G. We will denote this subgroup by $\Gamma_\pi G$ or ΓG if the action of π is clear from the context.

In applications we will be mainly concerned with two cases. Case I is when π acts on itself by inner automorphisms. Case II is when π acts on an abelian group M. When π acts on π by inner automorphisms we have $\Gamma^1(\pi) = \pi$, $\Gamma^\alpha(\pi) = [\pi, \Gamma^{\alpha-1}(\pi)]$ if α is a nonlimiting ordinal, and $\Gamma^\alpha(\pi) = \bigcap_{\beta < \alpha} \Gamma^\beta(\pi)$ if α is a limiting ordinal. When M is a π-module, $\Gamma^1(M) = M$, $\Gamma^\alpha(M) = (I\pi)\Gamma^{\alpha-1}(M)$ when α is a nonlimiting ordinal, and $\Gamma^\alpha(M) = {}_{\beta < \alpha}\Gamma^\beta(M)$ if α is a limiting ordinal.

We first deal with a slight extension of Stallings' theorem (Theorem 1.2).

Theorem 2.6

Let $f: G \twoheadrightarrow G'$ be an epimorphism of groups satisfying

$$H_1(f): H_1(G) \simeq H_1(G'),$$

$$H_2(f): H_2(G) \twoheadrightarrow H_2(G').$$

Then for any ordinal α, $G/\Gamma^\alpha(G) \xrightarrow{\bar{f}^\alpha} G'/\Gamma^\alpha(G')$ induced by f is an isomorphism.

Proof. By induction. Clearly $\bar{f}^\alpha$ is an isomorphism for $\alpha = 1$.   □

Assume $\bar{f}^\beta$ is an isomorphism $\forall$ ordinal $\beta < \alpha$.

Case I. α a nonlimiting ordinal. Then $\Gamma^\alpha(G) = [G, \Gamma^{\alpha-1}(G)]$ and

$$\bar{f}^{\alpha-1}: G/\Gamma^{\alpha-1}(G) \simeq G'/\Gamma^{\alpha-1}(G').$$

Associated to the diagram

$$
\begin{array}{ccc}
G & \twoheadrightarrow & G/\Gamma^{\alpha-1}(G) \\
\downarrow{\scriptstyle f} & & \downarrow{\scriptstyle \bar{f}^{\alpha-1}} \\
G' & \twoheadrightarrow & G'/\Gamma^{\alpha-1}(G')
\end{array} \quad ,
$$

$\exists$ a commutative diagram

$$
\begin{array}{ccccccccccc}
H_2(G) & \to & H_2(G/\Gamma^{\alpha-1}(G)) & \to & \Gamma^{\alpha-1}(G)/\Gamma^\alpha(G) & \to & H_1(G) & \to & H_1(G/\Gamma^{\alpha-1}(G)) & \to & 0 \\
\downarrow & & \downarrow & & \downarrow & & \downarrow & & \downarrow & & \\
H_2(G') & \to & H_2(G'/\Gamma^{\alpha-1}(G')) & \to & \Gamma^{\alpha-1}(G')/\Gamma^\alpha(G') & \to & H_1(G') & \to & H_1(G'/\Gamma^{\alpha-1}(G')) & \to & 0
\end{array}
$$

with exact rows. Hence $\Gamma^{\alpha-1}(G)/\Gamma^\alpha(G) \simeq \Gamma^{\alpha-1}(G')/\Gamma^\alpha(G')$.

It follows from the commutative diagram

$$
\begin{array}{ccccccccc}
1 & \to & \Gamma^{\alpha-1}(G)/\Gamma^\alpha(G) & \to & G/\Gamma^\alpha(G) & \to & G/\Gamma^{\alpha-1}(G) & \to & 1 \\
& & \downarrow & & \downarrow{\scriptstyle \bar{f}^\alpha} & & \simeq\downarrow{\scriptstyle \bar{f}^{\alpha-1}} & & \\
1 & \to & \Gamma^{\alpha-1}(G')/\Gamma^\alpha(G') & \to & G'/\Gamma^\alpha(G') & \to & G'/\Gamma^{\alpha-1}(G') & \to & 1
\end{array}
$$

that $\bar{f}^\alpha: G/\Gamma^\alpha(G) \simeq G'/\Gamma^\alpha(G')$.

Case II. α a limiting ordinal. Then $\Gamma^\alpha(G) = \bigcap_{\beta < \alpha} \Gamma^\beta(G)$. From $\bar{f}^\beta$ iso-morphism for every $\beta < \alpha$, it is immediate that $\bar{f}^\alpha: G/\Gamma^\alpha(G) \to G'/\Gamma^\alpha(G')$ is monic. But $f: G \twoheadrightarrow G'$ implies $\bar{f}^\alpha$ epic. Hence $\bar{f}^\alpha$ is an isomorphism.

Let G be a π group and $i: G \to \hat{G}$ the canonical map of G into the π completion of G. In case G is abelian (i.e., G is a π-module) then $\hat{G}$ is also an abelian group. We define $\Gamma'_\infty G = \operatorname{coker} i = \hat{G}/i(G)$. In case $G = \pi$ with π acting by inner automorphisms on itself we set $\Gamma'_\infty G$ equal to the left coset space $\hat{G}/i(G)$. There is a distinguished element in $\Gamma'_\infty G$, namely the coset $i(G)$.

Let G be a π group, G' a π' group, $\alpha: \pi \to \pi'$ a homomorphism, and $f: G \to G'$ be α-compatible. Then the map $\hat{f}: \hat{G} \to \hat{G}'$ described earlier (after Definition 2.4) yields a map $\Gamma'_\infty f: \Gamma'_\infty G \to \Gamma'_\infty G'$, which is a map of cosets if $G = \pi$, $G' = \pi'$, and the actions are respective inner automorphisms, or a map of abelian groups in case G is a π-module and G' is a π'-module. With these preliminaries we are now ready to state Dror's theorem.

Theorem 2.7 (E. Dror)

Let X, Y be 0-connected spaces and $f: X \to Y$ a map with $H_*(f): H_*(X) \simeq H_*(Y)$.

$\pi_1(X)$ acts on each $\pi_n(X)$ and $\pi_1(Y)$ acts on each $\pi_n(Y)$. The action of $\pi_1(X)$ [resp. $\pi_1(Y)$] on $\pi_1(X)$ [resp. $\pi_1(Y)$] is by inner automorphism. Thus we can talk of

$$\Gamma_\infty \pi_* f: \Gamma_\infty \pi_*(X) \to \Gamma_\infty \pi_*(Y),$$

$$\Gamma'_\infty \pi_* f: \Gamma'_\infty \pi_*(X) \to \Gamma'_\infty \pi_*(Y),$$

$$\Gamma \pi_* f: \Gamma \pi_*(X) \to \Gamma \pi_*(Y).$$

[observe that $f_*: \pi_*(X) \to \pi_*(Y)$ is a $\pi_1(f): \pi_1(X) \to \pi_1(Y)$ compatible map.]
If $f_*: H_*(X) \simeq H_*(Y)$, then $\hat{f}_*: \hat{\pi}_*(X) \simeq \hat{\pi}_*(Y)$. If further

(i) $\Gamma_\infty \pi_*(f): \Gamma_\infty \pi_*(X) \twoheadrightarrow \Gamma_\infty \pi_*(Y),$
(ii) $\Gamma'_\infty \pi_*(f): \Gamma'_\infty \pi_*(X) \twoheadrightarrow \Gamma'_\infty \pi_*(Y),$
(iii) $\Gamma \pi_*(f): \Gamma \pi_*(X) \twoheadrightarrow \Gamma \pi_*(Y),$

then $\pi_*(f): \pi_*(X) \simeq \pi_*(Y)$.

The preceding theorem is an immediate consequence of the following theorem.

Theorem 2.8

Let $f: X \to Y$ be a map of 0-connected spaces and n an integer greater than or equal to 1. Suppose $\pi_j(f): \pi_j(X) \simeq \pi_j(Y)$ for $j \leq n - 1$. Suppose further that

$$H_n(f): H_n(X) \simeq H_n(Y),$$

$$H_{n+1}(f): H_{n+1}(X) \twoheadrightarrow H_{n+1}(Y).$$

Then $\hat{\pi}_n(f)$: $\hat{\pi}_n(X) \simeq \hat{\pi}_n(Y)$. *If moreover*

(i) $\Gamma_\infty \pi_n(f)$: $\Gamma_\infty \pi_n(X) \twoheadrightarrow \Gamma_\infty \pi_n(Y)$,
(ii) $\Gamma'_\infty \pi_n(f)$: $\Gamma'_\infty \pi_n(X) \twoheadrightarrow \Gamma'_\infty \pi_n(Y)$,
(iii) $\Gamma \pi_n(f)$: $\Gamma \pi_n(X) \twoheadrightarrow \Gamma_\infty \pi_n(Y)$,

then $\pi_n(f)$: $\pi_n(X) \simeq \pi_n(Y)$.

Here $\pi_j(f)$: $\pi_j(X) \to \pi_j(Y)$ denotes the homomorphism induced by f. For proving Theorem 2.8 we need the following propositions and lemmas.

Proposition 2.9

Let $f: X \to Y$ *be a map of 0-connected spaces and* n *an integer greater than or equal to 1. Suppose* $\pi_j(f)$: $\pi_j(X) \simeq \pi_j(Y)$ *for* $0 \leq j \leq n - 1$,

$$H_n(f): H_n(X) \simeq H_n(Y),$$

$$H_{n+1}(f): H_{n+1}(X) \to H_{n+1}(Y).$$

Then

(i) *If* $n = 1$, *we get*

$$H_1(f): H_1(f): H_1(\pi_1(X)) \simeq H_1(\pi(Y)),$$

$$H_2(f): H_2(\pi_1(X)) \twoheadrightarrow H_2(\pi_1(Y)).$$

(ii) *If* $n > 1$, *we get*

$$H_0(\pi_1 X; \pi_n X) \overset{f_*}{\simeq} H_0(\pi_1 Y; \pi_n Y),$$

$$H_1(\pi_1 X; \pi_n X) \overset{f_*}{\twoheadrightarrow} H_1(\pi_1 Y; \pi_n Y).$$

Proof. Let us first deal with the case $n = 1$. Then

$$H_1(X) = \pi_1(X) | [\pi_1(X), \pi_1(X)] = H_1(\pi_1(X)) \overset{f_*}{\underset{\simeq}{\longrightarrow}} H_1(Y) = H_1(\pi_1(Y)).$$

Also by Hopf's theorem $H_2(\pi_1(X)) \simeq H_2(X)/\Sigma_2(X)$ where $\Sigma_2(X)$ is the subgroup of "spherical" homology classes in $H_2(X)$, namely $\Sigma_2(X)$ is the image of the Hurewicz homomorphism $\pi_2(X) \to H_2(X)$. Clearly $H_2(f)(\Sigma_2(X)) \subset \Sigma_2(Y)$ and hence $H_2(f)$ induces a map $H_2(X)/\Sigma_2(X) \to$

$H_2(Y)/\Sigma_2(Y)$. This is onto since $H_2(f) = H_2(X) \twoheadrightarrow H_2(Y)$. However $H_2(\pi_1(x)) \xrightarrow{H_2(f)} H_2(\pi_1(Y))$ is the same as the map $H_2(X)/\Sigma_2(X) \to H_2(Y)/\Sigma_2(Y)$ induced by f, when we identify $H_2(\pi_1(X))$ with $H_2(X)/\Sigma_2(X)$, etc.

Let $n > 1$. Then $\pi_1(f): \pi_1(X) \simeq \pi_1(Y)$. Using this isomorphism if we identify $\pi_1(X)$ and $\pi_1(Y)$, then $\pi_n(f): \pi_n(X) \to \pi_n(Y)$ is a π-module homomorphism where $\pi = \pi_1(X) = \pi_1(Y)$.

Let $P_n X \xrightarrow{\rho_n} P_{n-1} X$ be the nth stage of the Postnikov system for X with fiber $K(\pi_n(X), n)$ and similarly $P_n Y \xrightarrow{\rho_n'} P_{n-1} Y$ the nth stage Postnikov pair for Y with fiber $K(\pi_n(Y), n)$. f induces maps $f_n: P_n X \to P_n Y$ and $f_{n-1}: P_{n-1} X \to P_{n-1} Y$ with

$$
\begin{array}{ccc}
P_n X & \xrightarrow{f_n} & P_n Y \\
\downarrow{\scriptstyle \rho_n} & & \downarrow{\scriptstyle \rho_n'} \\
P_{n-1} X & \xrightarrow{f_{n-1}} & P_{n-1} Y
\end{array}
$$

homotopy commutative (actually f_n could be chosen to make the diagram commutative because ρ_n' is a fibration). In our case $\pi_j(f_{n-1}): \pi_j(P_{n-1}X) \to \pi_j(P_{n-1}Y)$ are isomorphisms for all j [because $\pi_i(P_{n-1}X) = 0 = \pi_i(P_{n-1}Y)$ for $i \geq n$]. Thus $f_{n-1}: P_{n-1}X \to P_{n-1}Y$ is a weak homotopy equivalence and hence induces isomorphisms in homology.

Associated to the fibration $\rho_n: P_n X \to P_{n-1} X$ there is a spectral sequence with $E_{p,q}^2 = H_p(P_{n-1}X; H_q(K(\pi_n(X), n)))$ and $E_{p,q}^\infty$ equal to the associated graded group corresponding to a nice filtration on $H_*(P_n X)$. Using this spectral sequence we will get an exact sequence

$$
H_{n+2}(P_n X) \to H_{n+2}(P_{n-1}X) \to H_1(\pi_1(X); P_n(X)) \to H_{n+1}(P_n X)
$$
$$
\to H_{n+1}(P_{n-1}X) \to H_0(\pi_1(X); \pi_n(X)) \to H_n(X)
$$
$$
\to H_n(P_{n-1}X) \to 0.
$$

For $n > 1$ one knows that $H_{n+1}(K(\pi_n(X), n)) = 0$. Also $H_q(K(\pi_n(X), n)) = 0$ for $0 < q < n$. Thus $E_{p,q}^2 = 0$ for $0 < q < n$ and $q = n + 1$. It follows that

$$
\boxed{E_{p,q}^r = 0 \quad \text{for } 0 < q < n \text{ and } q = n + 1 \text{ for all } r \geq 2} \tag{1}
$$

and since the spectral sequence is a strongly convergent one,

$$
\boxed{E_{p,q}^\infty = 0 \quad \text{for } 0 < q < n \text{ and } q = n + 1.} \tag{2}
$$

From (1) we see that $E^r_{n+2-r, r-1} = 0$ for $2 \leq r \leq n$. Also $E^r_{n+2+r, 1-r} = 0$ for $2 \leq r$ since $1 - r$ is negative then. It follows that

$$E^2_{n+2, 0} = E^3_{n+2, 0} = \cdots = E^{n+1}_{n+2, 0}. \tag{3}$$

Also from

$$0 = E_{2n+3, -n} \xrightarrow{d^{n+1}} E^{n+1}_{n+2, 0} \xrightarrow{d^{n+1}} E^{n+1}_{1, n},$$

we get

$$E^{n+2}_{n+2, 0} = \ker d^{n+1} : E^{n+1}_{n+2, 0} \to E^{n+1}_{1, n}.$$

Also $E^r_{n+2-r, r-1} = 0$ for $r \geq n + 3$ since $n + 2 - r$ is negative then. When $r = n + 2$, $E^{n+2}_{0, n+1, 1-r} = 0$ by the latter part of (1). Moreover $E^r_{n+2+r, 1-r} = 0$ for $r \geq n + 1$ since $1 - r$ is negative then. It follows that

$$E^{n+2}_{n+2, 0} = E^{n+3}_{n+2, 0} = \cdots = E^{\infty}_{n+2, 0}.$$

Thus we have

$$E^{\infty}_{n+2, 0} = E^{n+2}_{n+2, 0} = \ker d^{n+1} : E^{n+1}_{n+2, 0} \to E^{n+1}_{1, n}. \tag{4}$$

But

$$E^{\infty}_{n+2, 0} = \text{a quotient of } F_{n+2} H_{n+2}(P_n X)$$

$$= \text{a quotient of } H_{n+2}(P_n X).$$

Taking this surjective map $H_{n+2}(P_n X) \to E^{\infty}_{n+2, 0}$ and using (4) and (3) we get the exact sequence

$$H_{n+2}(P_n X) \longrightarrow E^2_{n+2, 0} \xrightarrow{d^{n+1}} E^{n+1}_{1, n} \longrightarrow E^{n+1}_{1, n} / \text{Im } d^{n+1} \longrightarrow 0. \tag{5}$$
$$\| \quad E^{n+1}_{n+2, 0}$$

Realizing that $E^2_{n+2, 0} = H_{n+2}(P_{n-1} X)$ [since $\pi_1(P_{n-1} X)$ acts trivially on $H_0(K(\pi_n(X), n)) = Z$] and that $E^{n+1}_{-n, 0} = 0$ we get from (5) the exact sequence

$$H_{n+2}(P_n X) \to H_{n+2}(P_{n-1} X) \to E^{n+1}_{1, n} \to E^{n+2}_{1, n} \to 0. \tag{5'}$$

From $E^r_{1+r, n+1-r} = 0 = E^r_{1-r, n-1+r}$ for $r \geq n + 2$ we get

$$E^{n+2}_{1, n} = E^{n+3}_{1, n} = \cdots = E^{\infty}_{1, n}.$$

Thus (5′) yields the exact sequence

$$H_{n+2}(P_n X) \to H_{n+2}(P_{n-1}X) \to E_{1,n}^{n+1} \to E_{1,n}^{\infty} \to 0. \qquad (5'')$$

Now, $E_{1,n}^{\infty} = F_1 H_{n+1}(P_n X)/F_0 H_{n+1}(P_n X)$. We know that $E_{2,n-1}^{\infty} = \cdots = E_{n,1}^{\infty} = 0$ and $E_{0,n+1}^{\infty} = 0$ from (2). Hence $F_1 H_{n+1}(P_n X) = \cdots = F_n H_{n+1}(P_n X)$ and $F_0 H_{n+1}(P_n X) = F_{-1}H_{n+1}(P_n X)$. But $F_{-1}H_{n+1}(P_n X) = 0$. Hence $F_0 H_{n+1}(P_n X) = 0$. It follows that $E_{1,n}^{\infty} = F_1 H_{n+1}(P_n X) \subset H_{n+1}(P_n X)$. But since $F_1 H_{n+1}(P_n X) = \cdots = F_n H_{n+1}(P_n X)$ we see that $E_{1,n}^{\infty} = F_n H_{n+1}(P_n X) \subset H_{n+1}(P_n X)$. But $H_{n+1}(P_n X)/F_n H_{n+1}(P_n X) = E_{n+1,0}^{\infty}$. From (5″) we now get the exact sequence

$$H_{n+2}(P_n X) \to H_{n+2}(P_{n-1}X) \to E_{1,n}^{n+1} \to H_{n+1}(P_n X) \to E_{n+1,0}^{\infty} \to 0. \quad (6)$$

Using the fact that $E_{p,q}^r = 0$ for $0 < q < n$ and $r \geq 2$ [see (1)] we immediately get $E_{n+1,0}^2 = \cdots = E_{n+1,0}^{n+1}$. From $E_{2n+2,-n}^{n+1} = 0$ we get $E_{n+1,0}^{n+2} = \ker d^{n+1}: E_{n+1,0}^{n+1} \to E_{0,n}^{n+1}$. From $E_{n+1-r,r-1}^r = 0 = E_{n+1-r,r-1}^r$ for $r \geq n+2$ we get $E_{n+1,0}^{n+2} = E_{n+1,0}^{n+3} = \cdots = E_{n+1,0}^{\infty}$. Thus $E_{n+1,0}^{\infty} = \ker d^{n+1}: E_{n+1,0}^{n+1} \to E_{0,n}^{n+1}$. Sequence (6) now yields the following exact sequence

$$H_{n+2}(P_n X) \longrightarrow H_{n+1}(P_{n-1}X) \longrightarrow E_{1,n}^{n+1} \longrightarrow H_{n+1}(P_n X)$$

$$\longrightarrow E_{n+1,0}^{n+1} \xrightarrow{d^{n+1}} E_{0,n}^{n+1} \text{ coker } d^{n+1} \longrightarrow 0). \qquad (7)$$

$$\big\| $$

$$E_{n+1,0}^2$$

Since $E_{n+1,0}^2 = H_{n+1}(P_{n-1}X)$, the preceding yields the exact sequence

$$H_{n+2}(P_n X) \to H_{n+2}(P_{n-1}X) \to E_{1,n}^{n+1} \to H_{n+1}(P_n X)$$

$$\to E_{0,n}^{n+1} \to E_{0,n}^{n+1}/\text{Im } d^{n+1} \to 0. \qquad (7')$$

From $E_{-n-1,2n}^{n+1} = 0$ we get $E_{0,n}^{n+2} = E_{0,n}^{n+1}/\text{Im } d^{n+1}$. Also $E_{-r,n+r-1}^r = 0 = E_{r,n+1-r}^r$ for $r \geq n+2$ yields $E_{0,n}^{n+2}E_{0,n}^{n+3} = \cdots = E_{0,n}^{\infty} = F_0 H_n(P_n X) \subset H_n(P_n X)$. From $E_{1,n-1}^{\infty} = \cdots = E_{n-1,1}^{\infty} = 0$ [see (2)] we get

$$F_0 H_n(P_n X) = F_1 H_n(P_n X) = \cdots = F_{n-1}H_n(P_n X).$$

Hence $H_n(P_n X)/E_{0,n}^{n+2} = [H_n(P_n X)]/[F_{n-1}H_n(P_n X)] = E_{n,0}^{\infty}$. Now (7′) yields the following exact sequence:

$$H_{n+2}(P_n X) \to H_{n+2}(P_{n-1}X) \to E_{1,n}^{n+1} \to H_{n+1}(P_n X) \to H_{n+1}(P_{n-1}X)$$

$$\to E_{0,n}^{n+1} \to H_n(P_n X) \to E_{n,0}^{\infty} \to 0. \qquad (8)$$

$E_{n+r,1-r}^r = 0 = E_{n-r,r-1}^r$ for $2 \leq r \leq n$ (observe that $E_{p,q}^r = 0$ for $0 < q < n$) yields $E_{n,0}^2 = \cdots = E_{n,0}^{n+1}$. Also $E_{n+r,1-r}^r = 0 = E_{n-r,r-1}^r$ for $r \geq n+1$ ($E_{n+r,1-r}^r = 0$ since $1 - r < 0$ and $E_{n-r,r-1}^r = 0$ since $n - r < 0$) yields $E_{n,0}^{n+1} = \cdots = E_{n,0}^\infty$. Thus $E_{n,0}^\infty = E_{n,0}^2 = H_n(P_{n-1}X)$. Substituting this in (8) we get the exact sequence:

$$H_{n+2}(P_nX) \to H_{n+2}(P_{n-1}X) \to E_{1,n}^{n+1} \to H_{n+1}(P_nX) \to H_{n+1}(P_{n-1}X)$$
$$\to E_{0,n}^{n+1} \to H_n(X) \to H_n(P_{n-1}X) \to 0. \tag{8'}$$

Observe that $H_i(P_nX) = H_i(X)$ for $i \leq n$.

From $E_{1-r,n+r-1}^r = 0 = E_{1+r,1+n-r}^r$ for $2 \leq r \leq n$ we get $E_{1,n}^2 = E_{1,n}^{n+1}$. Similarly $E_{-r,n-1}^r = 0 = E_{r,n+1-r}^r$ for $2 \leq r \leq n$ implies $E_{0,n}^2 = E_{0,n}^{n+1}$. But $E_{1,n}^2 = H_1(P_{n-1}X; \pi_nX)$, $E_{0,n}^2 = H_0(P_{n-1}X; \pi_nX)$.

Given any 0-connected space V and any local system A over V we have

$$H_0(V; A) \simeq H_0(\pi_1(V); A),$$
$$H_1(V; A) \simeq H_1(\pi_1(V); A),$$

where A is regarded as a right $\pi_1(V)$-module. This is because we can get a $K(\pi_1(V), 1)$ space as $V \cup$ (cells of dimension $n \geq 3$).

Substituting in (8') we get the exact sequence

$$H_{n+2}(P_nX) \to H_{n+2}(P_{n-1}X) \to H_1(\pi_1(X); \pi_n(X))$$
$$\to H_{n+1}(P_nX) \to H_{n+1}(P_{n-1}X) \to H_0(\pi_1(X); \pi_n(X))$$
$$\to H_n(X) \to H_n(P_{n-1}X) \to 0. \tag{9}$$

We will now complete the proof of Proposition 2.9 when $n > 1$. Naturality of the exact sequence (9) yields the commutative diagram

$$
\begin{array}{ccccccccc}
H_{n+2}(P_nX) & \to & H_{n+2}(P_{n-1}X) & \to & H_1(\pi_1(X);\pi_n(X)) & \to & H_{n+1}(P_nX) & \to & H_{n+1}(P_{n-1}X) \\
\downarrow{\scriptstyle f_{n*}} & & \downarrow{\scriptstyle f_{n-1*}} & & \downarrow{\scriptstyle H_1(f_*)} & & \downarrow{\scriptstyle f_{n*}} & & \downarrow{\scriptstyle f_{n-1*}} \\
H_{n+2}(P_nY) & \to & H_{n+2}(P_{n-1}Y) & \to & H_1(\pi_1(Y);\pi_n(Y)) & \to & H_{n+1}(P_nY) & \to & H_{n+1}(P_{n-1}Y) \\
& & & & \to H_0(\pi_1(X);\pi_n(X)) \to & & H_n(X) & \to H_n(P_{n-1}X) \to 0 \\
& & & & \downarrow{\scriptstyle H_0(f_*)} & & \downarrow{\scriptstyle f_*} & & \downarrow{\scriptstyle f_{n-1*}} \\
& & & & \to H_0(\pi_1(Y):\pi_n(Y)) \to & & H_n(Y) & \to H_n(P_{n-1}Y) \to 0
\end{array}
$$

Diagram A.

$f_{n-1}: P_{n-1}X \to P_{n-1}Y$ is a homotopy equivalence. Hence f_{n-1*}'s are all isomorphisms. By assumption $H_n(X) \overset{f_*}{\simeq} H_n(Y)$ and $H_{n+1}(X) \overset{f_*}{\longrightarrow} H_{n+1}(Y)$.

The last two terms in case of $P_{n+1}X \to P_n X$ and $P_{n+1}Y \to P_n Y$ of the analogue of exact sequence (9) yield the commutative diagram

$$
\begin{array}{ccc}
H_{n+1}(X) & \longrightarrow H_{n+1}(P_n X) & \longrightarrow 0 \\
\downarrow & \downarrow & \\
H_{n+1}(Y) & \longrightarrow H_{n+1}(P_n Y) & \longrightarrow 0
\end{array}
$$

with exact rows.

Hence $H_{n+1}(P_n X) \to H_{n+1}(P_n Y)$ is onto. Diagram A now implies $H_0(f_*)$: $H_0(\pi_1(X), \pi_n(X)) \simeq H_0(\pi_1(Y) : \pi_n(Y))$. Also from Diagram A we see that

$$
H_1(f_*): H_1(\pi_1(X); \pi_n(X)) \simeq H_1(\pi_1(Y); \pi_n(Y)).
$$

This completes the proof of Proposition 2.9. $\square$

Lemma 2.10

Let $f: A \to A'$ be a map of π-modules $B \subset (I\pi)A$ and $B' \subset (I\pi)A'$ with $f(B) \subset B'$. Suppose $H_1(\pi; f)$: $H_1(\pi; A) \twoheadrightarrow H_1(\pi; A')$ and

$$
A/B \xrightarrow[\simeq]{\bar{f}} A'/B'.
$$

Then f induces an isomorphism

$$
A/(I\pi)B \simeq A'/(I\pi)B'.
$$

Proof. The commutative diagram

$$
\begin{array}{ccccccccc}
0 & \longrightarrow & B & \longrightarrow & A & \longrightarrow & A/B & \longrightarrow & 0 \\
& & f\downarrow & & \downarrow f & & \downarrow \bar{f} & & \\
0 & \longrightarrow & B' & \longrightarrow & A' & \longrightarrow & A'/B' & \longrightarrow & 0
\end{array}
$$

of π-module homomorphisms, with the two rows exact yields the commutative diagram

$$
\begin{array}{ccccccccccc}
H_1(\pi; A) & \longrightarrow & H_1(\pi; A/B) & \xrightarrow{\partial} & H_0(\pi; B) & \longrightarrow & H_0(\pi; A) & \longrightarrow & H_0(\pi; A/B) & \longrightarrow & 0 \\
\downarrow f_* & & \downarrow \bar{f}_* & & \downarrow f_* & & \downarrow f_* & & \downarrow \bar{f}_* & & \\
H_1(\pi; A') & \longrightarrow & H_1(\pi; A'/B') & \xrightarrow{\partial} & H_0(\pi; B') & \longrightarrow & H_0(\pi; A') & \longrightarrow & H_0(\pi; A'/B') & \longrightarrow & 0
\end{array}
$$

with exact rows.

Also

$$H_0(\pi; A) = \frac{A}{(I\pi)A},$$

$$H_0\left(\pi; \frac{A}{B}\right) = \frac{A/B}{I\pi(A/B)} = \frac{A}{(I\pi)A + B} \simeq \frac{A}{(I\pi)A}$$

[since $B \subset (I\pi)A$]. The map $H_0(\pi; A) \to H_0(\pi; A/B)$ is thus an isomorphism. Similarly $H_0(\pi; A') \to H_0(\pi; A'/B')$ is also an isomorphism. It follows that

$$\begin{array}{ccccccc}
H_1(\pi; A) & \longrightarrow & H_1(\pi; A/B) & \stackrel{\partial}{\longrightarrow} & B/(I\pi)B & \longrightarrow & 0 \\
f_* \downarrow & & \simeq \downarrow \bar{f}_* & & \downarrow f_* & & \\
H_1(\pi; A') & \longrightarrow & H_1(\pi; A'/B') & \longrightarrow & B'/(I\pi)B' & \longrightarrow & 0
\end{array}$$

is commutative with exact rows and left extreme map onto and the next map $\bar{f}_*$ an isomorphism. It follows that

$$f_*: B/(I\pi)B \to B'/(I\pi)B' \quad \text{is an isomorphism.}$$

From the commutative diagram with exact rows

$$\begin{array}{ccccc}
B/(I\pi)B & \rightarrowtail & A/(I\pi)B & \longtwoheadrightarrow & A/B \\
\downarrow & & \downarrow \bar{f} & & \downarrow \bar{f} \\
B'/(I\pi)B' & \rightarrowtail & A'/(I\pi)B' & \longtwoheadrightarrow & A'/(I\pi)B'
\end{array} \quad ,$$

we see that $A/(I\pi)B \xrightarrow[\simeq]{\bar{f}} A'/(I\pi)B'$.

Lemma 2.11

Let $f: A \to A'$ be a map of π-modules. Suppose

$$H_0(\pi; f): H_0(\pi; A) \simeq H_0(\pi; A'),$$

$$H_1(\pi; f): H_1(\pi; A) \twoheadrightarrow H_1(\pi; A').$$

Then for all $k \geq 1$, $A/(I\pi)^k A \xrightarrow[\simeq]{\bar{f}} A'/(I\pi)^k A'$. Hence $\hat{f}: \hat{A} \simeq \hat{A}'$.

Proof. $H_0(\pi; A) = A/(I\pi)A$ and $H_0(\pi; A') = A'/(I\pi)A'$. From $H_0(\pi; f): H_0(\pi; A) \simeq H_0(\pi; A')$ we see that Lemma 2.11 is valid for $k = 1$.

Assume $k > 1$ and Lemma 2.11 valid for $k - 1$. Then $A/(I\pi)^{k-1}A \simeq A'/(I\pi)^{k-1}A'$. Take $B = (I\pi)^{k-1}A$ and $B' = (I\pi)^{k-1}A'$ in Lemma 2.10 to get

$$A/(I\pi)^k A \simeq A'/(I\pi)^k A'. \qquad \square$$

Proof of Theorem 2.8

First we deal with the case $n = 1$. From (i) of Proposition 2.9 we see that

$$H_1(f): H_1(\pi_1(X)) \simeq H_1(\pi_1(Y)),$$

$$H_2(f): H_2(\pi_1(X)) \twoheadrightarrow H_2(\pi_1(Y)).$$

Hence by Stallings' theorem

$$H_1(X)/\Gamma^k(\pi_1(X)) \xrightarrow{\;\hat{f}_k\;} \pi_1(Y)/\Gamma^k(\pi_1(Y))$$

are all isomorphisms for $k \geq 1$. Hence

$$\hat{\pi}_1(X) \xrightarrow{\;\hat{f}\;} \hat{\pi}_1(Y)$$

is an isomorphism. Now assume (i)–(iii) of Theorem 2.8 are valid for $n = 1$. We have a commutative diagram

$$
\begin{array}{ccccccc}
 & & & & & & \Gamma'_\infty \pi_1(X) \\
 & & & & & & \| \\
\Gamma_\infty \pi_1(X) & \overset{\mu}{\twoheadrightarrow} & \pi_1(X) & \xrightarrow{\;i\;} & \hat{\pi}_1(X) & \xrightarrow{\eta}\!\!\twoheadrightarrow & \text{left coset space } \hat{\pi}_1(X)/i(\pi_1(X)) \\
{\scriptstyle \Gamma_\infty \pi_1(f)}\downarrow & & \downarrow{\scriptstyle f} & & \downarrow{\scriptstyle \hat{f}} & & \uparrow{\scriptstyle \Gamma'_\infty \pi_1(f)} \\
\Gamma_\infty \pi_1(X) & \overset{\mu}{\twoheadrightarrow} & \pi_1(Y) & \xrightarrow{\;i\;} & \hat{\pi}_1(Y) & \xrightarrow{\eta}\!\!\twoheadrightarrow & \text{left coset space } \hat{\pi}_1(Y)/i(\pi_1(Y)) \\
 & & & & & & \| \\
 & & & & & & \Gamma'_\infty \pi_1(Y)
\end{array}
$$

with rows exact [for the square at the right extreme this means $\eta^{-1}(*) = \operatorname{Im} i$]. Let $b \in \pi_1(Y)$. Then $\exists$ an $x \in \hat{\pi}_1(X)$ with $\hat{f}(x) = i(b)$. $\Gamma'_\infty \pi_1(f)(\eta(x)) = \eta\hat{f}(x) = \eta i(b)$ is the trivial element of $\Gamma'_\infty \pi_1(Y)$. Hence $\eta(x)$ is the trivial element of $\Gamma'_\infty \pi_1(X)$. Hence $x = i(c)$ for some $c \in \pi_1(X)$. Now, $i(f(c)) = \hat{f}i(c) = \hat{f}(x) = i(b)$. Hence $b^{-1}f(c) = \mu(d)$ with $d \in \Gamma_\infty \pi_1(Y)$. Since $\Gamma_\infty \pi_1(f)$ is onto, $\exists$ some $u \in \Gamma_\infty \pi_1(X)$ with $d = \Gamma_\infty \pi_1(f)(u)$. Hence $b^{-1}f(c) =$

$\mu\Gamma_\infty\pi_1(f)(u) = f\mu(u)$ or $f(c)f(\mu(u)^{-1}) = b$. Thus

$$b = f\left(c \cdot \mu(u)^{-1}\right) \in \text{Im } f. \qquad (*)$$

This proves that $f: \pi_1(X) \to \pi_1(Y)$ is onto. $\exists$ an ordinal α such that

$$\left.\begin{array}{l} \Gamma^\alpha\big(\pi_1(X)\big) = \Gamma^\gamma\big(\pi_1(X)\big) \\ \Gamma^\alpha\big(\pi_1(Y)\big) = \Gamma^\gamma\big(\pi_1(Y)\big) \end{array}\right\} \quad \text{for all } \gamma \geq \alpha.$$

Then from $\Gamma\pi_1(X) \subset \pi_1(X)$ we get $\Gamma^\beta_{\pi_1(X)}(\Gamma(\pi_1(X))) \subset \Gamma^\beta(\pi_1(X))$ for all ordinals β. But $\Gamma^\beta_{\pi_1(X)}(\Gamma\pi_1(X)) = \Gamma\pi_1(X)$. Thus $\Gamma\pi_1(X) \subset \Gamma^\alpha\pi_1(X)$. Also $\Gamma^2_{\pi_1(X)}(\Gamma^\alpha(\pi_1(X))) = [\pi_1(X), \Gamma^\alpha\pi_1(X)] = \Gamma^{\alpha+1}(\pi_1(X)) = \Gamma^\alpha(\pi_1(X))$. Thus $\Gamma^\alpha(\pi_1(X))$ is a $\pi_1(X)$-perfect subgroup of $\pi_1(X)$. Hence $\Gamma^\alpha(\pi_1(X)) \subset \Gamma\pi_1(X)$. It follows that $\Gamma\pi_1(X) = \Gamma^\alpha\pi_1(X)$. Similarly $\Gamma(\pi_1(Y)) = \Gamma^\alpha\pi_1(Y)$.

We have already seen that $f: \pi_1(X) \to \pi_1(Y)$ is onto [see $(*)$]. Thus by Theorem 2.6, $\bar{f}^\alpha: \pi_1(X)/\Gamma^\alpha\pi_1(X) \simeq \pi_1(Y)/\Gamma^\alpha\pi_1(Y)$.

From the following commutative diagram with exact rows

$$
\begin{array}{ccccc}
\Gamma\pi_1(X) = \Gamma^\alpha\pi_1(X) & \rightarrowtail & \pi_1(X) & \twoheadrightarrow & \pi_1(X)/\Gamma\pi_1(X) \\
\downarrow & & \downarrow f & & \downarrow \\
\Gamma\pi_1(Y) = \Gamma^\alpha\pi_1(Y) & \rightarrowtail & \pi_1(Y) & \twoheadrightarrow & \pi_1(Y)/\Gamma\pi_1(Y)
\end{array}
$$

we see that $f: \pi_1(X) \to \pi_1(Y)$ is monic.

$(*)$ and $(**)$ imply that $\pi_1(X) \xrightarrow{f} \pi_1(Y)$ is an isomorphism.

We now take up the case when $n \geq 2$. In this case $\pi_1(f): \pi_1(X) \simeq \pi_1(Y)$. Using this isomorphism we identify the two groups and write π for either of them. Then $\pi_n(f): \pi_n(X) \to \pi_n(Y)$ is a π homomorphism. By (ii) of Proposition 2.9 we see that

$$H_0\big(\pi; \pi_n(X)\big) \overset{\pi_n(f)_*}{\simeq} H_0\big(\pi; \pi_n(Y)\big),$$

$$H_1\big(\pi; \pi_n(X)\big) \xrightarrow{\pi_n(f)_*} H_1\big(\pi; \pi_n(Y)\big).$$

We will write f_* for the maps $\pi_n(f)_*$.

Lemma 2.11 then implies $\pi_n(X)/(I\pi^k)\pi_n(X) \simeq \pi_n(Y)/(I\pi)^k\pi_n(Y)$ for all $k \geq 1$. Hence $\hat{\pi}_n(X) \overset{\hat{f}}{\simeq} \hat{\pi}_n(Y)$.

Now, assume that conditions (i)–(iii) of Theorem 2.8 are satisfied. We have a commutative diagram

$$
\begin{array}{ccccccc}
\Gamma_\infty \pi_n(X) & \overset{\mu}{\twoheadrightarrow} & \pi_n(X) & \overset{i}{\longrightarrow} & \hat{\pi}_n(X) & \overset{\eta}{\twoheadrightarrow} & \hat{\pi}_n(Y)/i(\pi_n(Y)) = \Gamma'_\infty \pi_n(X) \\
{\scriptstyle \Gamma_\infty \pi_n(f)}\downarrow & & {\scriptstyle f}\downarrow & & \downarrow{\scriptstyle f} & & {\scriptstyle \Gamma'_\infty \pi_n(f)}\downarrow \\
\Gamma_\infty \pi_n(Y) & \overset{\mu}{\twoheadrightarrow} & \pi_n(Y) & \overset{i}{\longrightarrow} & \hat{\pi}_n(Y) & \overset{\eta}{\twoheadrightarrow} & \hat{\pi}_n(Y)/i(\pi_n(Y)) = \Gamma'_\infty \pi_n(Y)
\end{array}
$$

with exact rows. It follows that $f: \pi_n(X) \to \pi_n(Y)$ is onto. Exactly as in the case $n = 1$, it can be shown that $\exists$ an ordinal α with

$$\Gamma \pi_n(X) = (I\pi)^\alpha \pi_n(X),$$

$$\Gamma \pi_n(Y) = (I\pi)^\alpha \pi_n(X),$$

where

$$(I\pi)^\alpha \pi_n(X) = (I\pi)(I\pi)^{\alpha-1} \pi_n(X) \quad \text{if } \alpha \text{ is nonlimiting}$$

$$= \bigcap_{\beta > \alpha} (I\pi)^\beta \pi_n(X) \quad \text{if } \alpha \text{ is limiting.}$$

We know that $f: \pi_n(X) \to \pi_n(Y)$ is onto.

Assume that we have shown that $\pi_n(X)/(I\pi)^\beta \pi_n(X) \simeq \pi_n(Y)/(I\pi)^\beta \pi_n(Y)$ for all $\beta < \alpha$. Of course for $\beta = 1$, this is the assertion $H_0(\pi; \pi_n(X)) \overset{f_*}{\simeq} H_0(\pi; \pi_n(Y))$. If α is nonlimiting, taking $A = \pi_n(X)$, $A' = \pi_n(Y)$, $B = (I\pi)^{\alpha-1}A$, and $B' = (I\pi)^{\alpha-1}A'$, we see from Lemma 2.10 that $A/(I\pi)^\alpha A \simeq A'/(I\pi)^\alpha A'$ or $\pi_n(X)/\Gamma\pi_n(X) \simeq \pi_n(Y)/\Gamma\pi_n(Y)$. If α is limiting, then from $A/(I\pi)^\beta A \simeq A'/(I\pi)^\beta A'$ for all $\beta < \alpha$, $(I\pi)^\alpha A = \bigcap_{\beta<\alpha}(I\pi)^\beta A$, and $(I\pi)^\alpha A' = \bigcap_{\beta<\alpha}(I\pi)^\beta A'$ we immediately see that $A/(I\pi)^\alpha A \twoheadrightarrow A'/(I\pi)^\alpha A'$ or equivalently

$$\pi_n(X)/\Gamma\pi_n(X) \twoheadrightarrow \pi_n(Y)/\Gamma\pi_n(Y).$$

However $\pi_n(X) \overset{f}{\twoheadrightarrow} \pi_n(Y)$. Hence $\pi_n(X)/\Gamma\pi_n(X) \simeq \pi_n(Y)/\Gamma\pi_n(Y)$.

Now from the commutative diagram

$$
\begin{array}{ccccc}
\Gamma\pi_n(X) & \twoheadrightarrow & \pi_n(X) & \longrightarrow & \hat{\pi}_n(X)/\Gamma\pi_n(X) \\
\downarrow & & \downarrow{\scriptstyle f} & & \downarrow \\
\Gamma\pi_n(Y) & \twoheadrightarrow & \pi_n(Y) & \longrightarrow & \hat{\pi}_n(Y)/\Gamma\pi_n(Y)
\end{array}
$$

with exact rows we see that $f: \pi_n(X) \to \pi_n(Y)$ is monic. Hence $f: \pi_n(X) \simeq \pi_n(Y)$.

This completes the proof of Theorem 2.8. $\square$

3   NILPOTENT SPACES

Definition 3.1

An arcwise-connected space X is said to be *nilpotent* if the usual action of $\pi_1(X)$ on $\pi_n(X)$ is nilpotent for all $n \geq 1$. This is equivalent to saying that $\pi_1(X)$ is nilpotent and for $n \geq 2$, $\pi_n(X)$ is a nilpotent $\pi_1(X)$-module.

For a nilpotent space X, we have $\Gamma^k_{\pi_1(X)}(\pi_n(X)) = \{1\}$ for some k and hence

$$\hat{\pi}_n(X) = \pi_n(X), \qquad \Gamma_\infty \pi_n(X) = \{1\}, \qquad \Gamma \pi_n(X) \subset \Gamma_\infty \pi_n(X) = \{1\},$$

and $\pi'_\infty \pi_n(X) = \{1\}$ since $\hat{\pi}_n(X) = \pi_n(X)$. Thus as a corollary of Theorem 2.8 we obtain the following result.

Theorem 3.2

Let $f: X \to Y$ be a map of 0-connected nilpotent spaces satisfying $f_*: H_*(X) \simeq H_*(Y)$. Then $f_*: \pi_*(X) \simeq \pi_*(Y)$.

If further X, Y are CW-complexes f is a homotopy equivalence.

Example 3.3

All simply connected spaces are trivially nilpotent. Thus from Theorem 3.2 we recover Whitehead's theorem.

Example 3.4

All H spaces are nilpotent spaces. Let X be a 0-connected H space. We will prove that the action of $\pi_1(X)$ on $\pi_n(X)$ is trivial for $n \geq 1$. In particular $\pi_1(X)$ itself is abelian. Recall that the action of $\pi_1(X)$ on $\pi_n(X)$ is defined as follows:

Let $\alpha \in \pi_n(X)$ and $a \in \pi_1(X)$. Let $f: (S^n, *) \to (X, x_0)$ be any map representing α and $\theta: (I, \dot{I}) \to (X, x_0)$ be any loop representing a. Then $\varphi: S^n \times 0 \cup * \times I \to X$ given by $\varphi(u, 0) = f(u) \varphi(*, t) = \theta(t)$ for every $u \in S^n$ and $t \in I$ admits of an extension $\Phi: S^n \times I \to X$. Then $g: S^n \to X$ given by $g(u) = \Phi(u, 1)$ represents $a \cdot \alpha$. It is known that $a \cdot \alpha$ depends only on a and α and not on the representatives f, θ nor on the extension Φ of φ.

Now let (X, x_0) be an H space, $\alpha \in \pi_n(X)$, and $a \in \pi_1(X)$. Let $f: S^n \to X$ represent α and $\theta: (I, \dot{I}) \to (X, x_0)$ represent a. Then $F: S^n \times I \to X$ given by $F(u, t) = \mu(f(u), \theta(t))$ where $\mu: (X \times X, x_0 \times x_0) \to (X, x_0)$ is the multiplication on X, satisfies the following conditions:

$$F(u, 0) = \mu(f(u), x_0),$$

$$F(*, t) = \mu(x_0, \theta(t)).$$

$f'(u) = F(u, 0)$ represents α and $\theta'(t) = F(*, t)$ $\mu(x_0, \theta(t))$ represents a. Hence $a \cdot \alpha \in \pi_n(X)$ is represented by $g'(u) = F(u, 1): (S^n, *) \to (X, x_0)$. But $g'(u) = \mu(f(u), x_0) = f'(u)$. Hence $a \cdot \alpha = \alpha$.

Example 3.5

Let n be any integer greater than or equal to 1 and $P^n\mathbb{R}$ the real projective space of dimension n. We will show that $P^n\mathbb{R}$ is a nilpotent space $\Leftrightarrow n$ is odd.

For $n = 1$, $P^1\mathbb{R}$ is homeomorphic to S^1 and is an H space. Thus $P^1\mathbb{R}$ is nilpotent. Let $n \geq 2$. Then $\pi_1(P^n\mathbb{R}) = Z/2Z$ and $S^n \xrightarrow{\eta} P^n\mathbb{R}$ is the universal covering of $P^n\mathbb{R}$, where η is the canonical quotient map.

$\pi_1(P^n\mathbb{R}) = Z/2Z$ acts as the deck transformation group. The nonzero element a of $\pi_1(P^n\mathbb{R})$ acts as the antipodal map $S^n \to S^n$, namely $a \cdot x = -x$ $\forall x \in S^n$.

If n is odd, the antipodal map has degree 1 and hence is homotopic to the identity of S^n. Hence $\pi_1(P^n\mathbb{R})$ acts trivially on $\pi_k(P^n\mathbb{R})$ for all $k \geq 2$. Already $\pi_1(P^n\mathbb{R})$ is abelian. Thus when n is odd, $P^n\mathbb{R}$ is nilpotent.

Suppose n is even. Then the antipodal map has degree -1. If $t \in \pi_n(S^n) \simeq Z$ generates $\pi_n(S^n)$, then $a \cdot t = -t$ in Z. Writing π for $\pi_1(P^n\mathbb{R})$, we see that π acts on $\pi_n(P^n\mathbb{R}) = Z$ by $a \cdot t = -t$. Then $\Gamma_\pi^k(\pi_n(P^n\mathbb{R})) \supset 2^k Z$ and hence this action of π on $\pi_n(P^n\mathbb{R})$ is not nilpotent.

4   NILPOTENCE OF SEMI-DIRECT PRODUCTS

Let π act on a group G via $\theta: \pi \to \text{Aut } G$ and let $H = G \times_\theta \pi$ be the semi-direct product. Recall that in H we have

$$(x_1, a_1)(x_2, a_2) = (x_1 \cdot \theta(a_1)(x_2), a_1 a_2)$$

for any x_1, x_2 in G, a_1, a_2 in π. It is clear that $\{(1, a) | a \in \pi\}$ is a subgroup of H isomorphic to π and $\{(x, 1) | x \in G\}$ is a subgroup of H isomorphic to G.

Also $(x, a)^{-1} = (\theta(a^{-1})(x^{-1}), a^{-1})$. It is easily seen that $G' = \{(x, 1)|x \in G\}$ is a normal subgroup of H with $H/G' \simeq \pi$.

Our aim is to prove the following.

Theorem 4.1

Let π act on the group G via $\theta: \pi \to \text{Aut } G$. Then $H = G \times_\theta \pi$ is nilpotent $\Leftrightarrow \pi$ is nilpotent and the action of π on G is nilpotent.

We need some lemmas for proving this result.

Lemma 4.2

With the notations as in Theorem 4.1 we have $\Gamma^i(H) \subset G \times \Gamma^i(\pi)$ [where $G \times \Gamma^i(\pi)$ denotes the subgroup of H consisting of elements of the form (x, a) with $x \in G$ and $a \in \Gamma^i(\pi)$] for every integer $i \geq 1$.

Proof. Straightforward induction on i. □

For any $i \geq 1$ let $L^i = \{x \in G | (x, 1) \in \Gamma^i(H)\}$. Alternatively, $L^i \times \{1\} = \Gamma^i(H) \cap (G \times \{1\})$.

Lemma 4.3

L^i is a normal π-subgroup of G.

Proof. Since $\Gamma^i(H)$ is a normal subgroup of H, it follows that $\Gamma^i(H) \cap G'$ is a normal subgroup of G' where $G' = \{(x, 1) \in H | x \in G\}$. Since $G \simeq G'$ under $x \to (x, 1)$ we see that L^i is a normal subgroup of G.

Let $x \in L^i$ and $a \in \pi$. Then $(x, 1) \in \Gamma^i(H)$ and $\Gamma^i(H)$ is a normal subgroup of H. Hence $(1, a)(x, 1)(1, a)^{-1} \in \Gamma^i(H)$. But

$$
\begin{aligned}
(1, a)(x, 1)(1, a)^{-1} &= (\theta(a)(x), a)(1, a^{-1}) \\
&= (\theta(a)(x) \cdot \theta(a)(1), 1\,1) \\
&= (\theta(a)(x), 1).
\end{aligned}
$$

Hence $\theta(a)(x) \in L^i$. This proves Lemma 4.3. □

Lemma 4.4

$L^i \supset \Gamma^i_\pi(G)$ for all $i \geq 1$.

Proof. From $\Gamma^1(H) = H$ we see that $L^1 = G$ and hence the lemma is trivially valid for $i = 1$. Assume the lemma is valid for i. Then $\Gamma^i(H) \supset$

$\Gamma_\pi^i(G) \times \{1\}$. Hence

$$\Gamma^{i+1}(H) = [H, \Gamma^i(H)] \supset [G \times \{1\}, \Gamma_\pi^i(G) \times \{1\}]$$
$$\supset [G, \Gamma_\pi^i(G)] \times \{1\}.$$

Hence $L^{i+1} \supset [G, \Gamma_\pi^i(G)]$.

Let $c \in \Gamma_\pi^i(G)$ and $a \in \pi$. Then $(c, 1) \in \Gamma^i(H)$. Hence

$$(1, a)(c, 1)(1, a)^{-1}(c, 1)^{-1} \in \Gamma^{i+1}(H).$$

However

$$(1, a)(c, 1)(1, a)^{-1}(c, 1)^{-1} = (\theta(a)(c), a)\,(1, a^{-1})(c^{-1}, 1)$$
$$= (\theta(a)(c), a)(\theta(a^{-1})(c^{-1}), a^{-1})$$
$$= (\theta(a)(c) \cdot \theta(a)(\theta(a^{-1})(c^{-1}), 1))$$
$$= (\theta(a)(c) \cdot c^{-1}, 1).$$

Thus $(\theta(a)(c) \cdot c^{-1}, 1) \in \Gamma^{i+1}(H)$ or $\theta(a)(c) \cdot c^{-1} \in L^{i+1}$.

From Lemma 4.3, L^{i+1} is a normal π subgroup of G. We have shown that $L^{i+1} \supset [G, \Gamma_\pi^i(G)] \cup \{\theta(a)(c) \cdot c^{-1} | c \in \Gamma_\pi^i(G), \ a \in \pi\}$. Hence $L^{i+1} \supset \Gamma_\pi^{i+1}(G)$. This proves Lemma 4.4. $\square$

Lemma 4.5

Suppose π is nilpotent with $\Gamma^k(\pi) = \{1\}$. Then $\Gamma^{k+i}(H) \subset \Gamma_\pi^{i+1}(G) \times \{1\}$ for all $i \geq 0$.

Proof. Lemma 4.2 yields $\Gamma^k(H) \subset G \times \Gamma^k(\pi) = G \times \{1\}$. Thus for $i = 0$ Lemma 4.5 is valid. Assume Lemma 4.5 is valid for i. Then any element of $\Gamma^{k+i}(H)$ is of the form $(x, 1)$ with $x \in \Gamma_\pi^{i+1}(G)$. Let $(y, a) \in H$ be arbitrary with $y \in G$ and $a \in \pi$. Then

$$(y, a)(x, 1)(y, a)^{-1}(x, 1)^{-1}$$
$$= (y \cdot \theta(a)(x), a)(\theta(a^{-1})(y^{-1}), a^{-1})(x^{-1}, 1)$$
$$= (y \cdot \theta(a)(x), a)(\theta(a^{-1})(y^{-1}) \cdot \theta(a^{-1})(x^{-1}), a^{-1})$$
$$= (y \cdot \theta(a)(x), a)(\theta(a^{-1})(y^{-1}x^{-1}), a^{-1})$$
$$= (y \cdot \theta(a)(x) \cdot \theta(a)(\theta(a^{-1})(y^{-1}x^{-1})), 1)$$
$$= (y\theta(a)(x) \cdot x^{-1}y^{-1}yxy^{-1}x^{-1}, 1).$$

From $x \in \Gamma_\pi^{i+1}(G)$ we see that $\theta(a)(x) \cdot x^{-1} \in \Gamma_\pi^{i+2}(G)$. Also $\Gamma_\pi^{i+2}(G)$ is normal in G. Hence $y\theta(a)(x)x^{-1}y^{-1} \in \Gamma_\pi^{i+2}(G)$. From $x \in G_\pi^{i+1}(G)$ we get $yxy^{-1}x^{-1} \in [G, \Gamma_\pi^{i+1}(G)] \subset \Gamma_\pi^{i+2}(G)$. Hence $y\theta(a)(x)x^{-1}y^{-1}yxy^{-1}x^{-1} \in \Gamma_\pi^{i+2}(G)$ or $y\theta(a)(x)y^{-1}x^{-1} \in \Gamma_\pi^{i+2}(G)$. It follows that $\Gamma^{k+i+1}(H) \subset \Gamma_\pi^{i+2}(G) \times \{1\}$. This proves Lemma 4.5. $\square$

Proof of Theorem 4.1

Let π be nilpotent and the action of π on G be nilpotent. Then $\exists$ integers $k \geq 1$ and $\ell \geq 1$ with $\Gamma^k(\pi) = \{1\}$ and $\Gamma_\pi^\ell(G) = \{1\}$. From Lemma 4.5 we get

$$\Gamma^{k+\ell}(H) \subset \Gamma_\pi^{\ell+1}(G) \times \{1\} = \{1\}.$$

Hence H is nilpotent.

Conversely, assume H is nilpotent. As π is a quotient group of H it follows that π is nilpotent. From Lemma 4.4 we see that $\Gamma^i(H) \supset \Gamma_\pi^i(G) \times \{1\}$ for all $i \geq 1$. If $\Gamma^\mu(H) = \{1\}$, then $\Gamma_\pi^\mu(G) = \{1\}$. Hence the action of π on G is nilpotent. This completes the proof of Theorem 4.1. The results in this section are essentially due to Hilton [8]. $\square$

5 NILPOTENCE OF CLASSIFYING SPACES

Let G be a locally arcwise-connected topological group and G_0 the identity component of G. Let B_G denote the classifying space of G. One knows that G_0 is a normal subgroup of G and that $\pi_0(G) = G/G_0$ is a group. Also for $n \geq 1$, $\pi_n(G, e) \simeq \pi_n(G_0, e)$. Moreover using the homotopy exact sequence of the fibration $E_G \to B_G$ with fiber G (where E_G is the total space of the universal bundle) we can identify

$$\pi_0(G) = G/G_0 \quad \text{with } \pi_1(B_G),$$
$$\pi_n(G, e) = \pi_n(G_0, e) \quad \text{with } \pi_{n+1}(B_G) \text{ for } n \geq 1.$$

The usual action of $\pi_1(B_G)$ on $\pi_{n+1}(B_G)$ gives rise to action of $\pi = \pi_0(G) = G/G_0$ on $\pi_n(G_0, e)$ when we identify $\pi_1(B_G)$ with $\pi_0(G)$ and $\pi_{n+1}(B_G)$ with $\pi_n(G_0, e)$. This action could also be described as follows:

For any $g \in G$ let Int $g : G \to G$ be the inner automorphism Int $g(x) = gxg^{-1}$. Then Int $g(G_0) = G_0$ and thus Int $g|G_0 : G_0 \to G_0$ is an automorphism of G_0. Thus there is an induced automorphism

$$g_\# : \pi_n(G_0) \to \pi_n(G_0).$$

We get an action of G on $\pi_n(G_0)$ by setting $H_n : G \to \text{Aut } \pi_n(G_0)$ to be the map $H_n(g) = g_\#$. When $g \in G_0$, the map Int $g : G_0 \to G_0$ is homotopic to the identity map. In fact if σ is a path in G_0 with $\sigma(0) = g$ and $\sigma(1) = e$, then

$G_0 \times I \stackrel{F}{\to} G_0$ given by $F(x, t) = \text{Int } \sigma(t)(x)$ is a homotopy between Int g and Id. Thus $g_\# = \text{Id}_{\pi_n(G_0)}$ whenever $g \in G_0$.

Hence we get an action of G/G_0 on $\pi_n(G_0)$ under $\theta_n(\eta(g)) = g_\# \in \text{Aut } \pi_n(G_0)$, where $\eta: G \to G/G_0$ is the canonical quotient map.

In particular B_G is nilpotent $\Leftrightarrow$ G/G_0 is an abstract nilpotent group and the action of G/G_0 on $\pi_n(G_0)$ just described is nilpotent.

Using Theorem 4.1 we immediately get the following:

Proposition 5.1

B_G is a nilpotent space $\Leftrightarrow$ all the semi-direct products $\pi_n(G_0) \times_{\theta_n} G/G_0$ are nilpotent for $n \geq 1$.

Let $[[S^n, G]]$ denote the set of "free homotopy classes" of maps of S^n in G. Using pointwise multiplication of maps one converts $[[S^n, G]]$ into a group.

G_0 being an arcwise-connected topological group, it follows that $\pi_1(G_0, e)$ acts trivially on all the $\pi_n(G, e)$'s. Because of this one can identify the group of free homotopy classes $[[S^n, G_0]]$ with $\pi_n(G_0)$.

Theorem 5.2

$[[S^n, G]] \simeq \pi_n(G_0) \times_{\theta_n} (G/G_0)$ for each $n \geq 1$.

Proof. Let $f: S^n \to G$ be any continuous map. Since S^n is arcwise-connected, $f(S^n)$ will lie in one connected component of G. Hence we can find an element $g_f \in G$ with $f(S^n) \subset G_0 g_f$. If g_f and g_f' are any two elements of G with $f(S^n) \subset G_0 g_f$ and $f(S^n) \subset G_0 g_f'$, then from $G_0 g_f = G_0 g_f'$ we see that g_f, g_f' lie in the same coset of G_0 in G. Hence $\eta(g_f) \in G/G_0$ depends only on f. Also if $f \sim f'$ as maps of S^n in G, since $S^n \times I$ is arcwise-connected we see that $f(S^n)$ and $f'(S^n)$ lie in the same coset of G_0 in G. Hence $\eta(g_f) = \eta(g_{f'})$ in G/G_0. Let $\bar{f}: S^n \to G_0$ be defined by $\bar{f}(u) = f(u)g_f^{-1}$. If g_f, g_f' are any two elements of G with $f(S^n) \subset G_0 g_f = G_0 g_f'$, then $h = g_f g_f'^{-1} \in G_0$. Thus $\bar{f}(u) = f(u)g_f'^{-1} = \bar{f}(u)h$ with $h \in G_0$. Picking a path in G_0 joining h to e we see that $\bar{f} \sim \bar{f}$ as maps of S^n in G_0. Hence the class $[\bar{f}]$ of $\bar{f}$ in $\pi_n(G_0)$ is independent of the choice of g_f. Thus $[[S^n, G]] \stackrel{\alpha}{\to} \pi_n(G_0) \times G/G_0$ defined by $[f] \mapsto ([\bar{f}], \eta(g_f))$ where $f(u) = \bar{f}(u)g_f$ with $\bar{f}(S^n) \subset G_0$ is a well-defined set theoretic map. It is easily seen that it is a set theoretic bijection.

Thus to complete the proof of Theorem 5.2 we have only to check that α is a group homomorphism of $[[S^n, G]]$ in $\pi_n(G_0) \times_{\theta_n} (G/G_0)$. Let $f_1: S^n \to G$ and $f_2: S^n \to G$ be any two maps. Then $[f_1] \cdot [f_2] = [f_1 \cdot f_2]$ where $(f_1 \cdot f_2)(u) = f_1(u) \cdot f_2(u) \ \forall \ u \in S^n$. Let g_1, g_2 in G be such that $f_1(S^n) \subset G_0 g_1$ and $f_2(S^n) \subset G_0 g_2$. Then

$$\left. \begin{aligned} \alpha([f_1]) &= ([\bar{f}_1], \eta(g_1)) \quad \text{where } \bar{f}_1(u) = f_1(u)g_1^{-1} \\ \alpha([f_2]) &= ([\bar{f}_2], \eta(g_2)) \quad \text{where } \bar{f}_2(u) = f_2(u)g_2^{-1} \end{aligned} \right\} \quad \forall \ u \in S^n.$$

Now

$$f_1(u)f_2(u) = \bar{f}_1(u)g_1\bar{f}_2(u)g_2$$
$$= \bar{f}_1(u)g_1\bar{f}_2(u)g_1^{-1} \cdot g_1g_2.$$

If $\varphi: S^n \to G$ is given by $\varphi(u) = \bar{f}_1(u) \cdot g_1\bar{f}_2(u)g_1^{-1}$, then $\varphi(S^n) \subset G_0$. Thus by definition of α,

$$\alpha([f_1 \cdot f_2]) = ([\varphi], \eta(g_1, g_2))$$
$$= ([\bar{f}_1]g_{1\#}[\bar{f}_2], \eta(g_1)\eta(g_2))$$
$$= ([\bar{g}_1]\theta_n(g_1)([\bar{f}_2]), \eta(g_1)\eta(g_2))$$
$$= ([\bar{f}_1], \eta(g_1))([\bar{f}_2], \eta(g_2)) \quad \text{in } \pi_n(G_0) \times_{\theta_n} (G)(G_0).$$

This completes the proof of Theorem 5.2. □

Corollary 5.3

The group $[[S^n, G]]$ is nilpotent for all $n \geq 1 \Leftrightarrow B_G$ is a nilpotent space.

Proposition 5.4

(i) For all odd n, $[[S^i, O(n)]]$ is a nilpotent group for each integer $i \geq 1$.

(ii) For $n \geq 1$, $[[S^{2n-1}, O(2n)]]$ is not nilpotent.

Proof. It is known that $O(n)/SO(n) \simeq Z/2Z$ acts trivially on $\pi_i(SO(n))$ for all $i \geq 1$, whenever n is odd [21]. This yields (i).

There exists an exact sequence $0 \to Z \to \pi_{2n-1}(SO(2n)) \twoheadrightarrow \pi_{2n-1}(SO(2n + 1))$ where on the generator t_{2n} of $Z \subset \pi_{2n-1}(SO(2n))$ the group $O(2n)/SO(2n) = Z/2Z$ acts by $\tau(t_{2n}) = -t_{2n}$ where $\tau \neq 0$ in $Z/2Z$. Hence the semi-direct product $\pi_{2n-1}(SO(2n)) \times_{\theta_{2n-1}} O(2n)/SO(2n)$ contains the infinite dihedral group as a subgroup. This yields (ii). □

Corollary 5.5

Let $n \geq 1$. The Grassmann space $G_{n,\infty}$ of n planes in $\mathbb{R}^\infty$ is nilpotent if and only if n is odd. This is because $G_{n,\infty} = BO(n)$.

The results in this section are essentially due to Roitberg [18].

Chapter Three

Grothendieck Groups

The rings we consider will be associative rings with $1 \neq 0$. All the modules we consider will be unitary modules. Unless otherwise mentioned we deal with left modules. For any ring R, let R-mod (resp. mod-R) denote the category of left R-modules (resp. right R-modules). For any module M we denote the isomorphism class of M by $\langle M \rangle$. All the ring homomorphisms we consider have to satisfy the condition $f(1) = 1$.

1 DEFINITION OF $K_0(R)$, $\tilde{K}_0(R)$, $G_0(R)$, $\tilde{G}_0(R)$, AND THEIR BASIC PROPERTIES

Let R be a ring. Let $\underline{P}(R)$ [resp. $\underline{M}(R)$] denote the full subcategory of R-mod consisting of finitely generated projective R-modules (resp. all finitely generated R-modules). Let $\mathscr{C}$ denote any one of $\underline{P}(R)$ or $\underline{M}(R)$. In fact for our initial considerations, $\mathscr{C}$ could be any full additive subcategory of R-mod with the property that the isomorphism classes of objects of $\mathscr{C}$ form a set. Let $\langle \text{obj}\, \mathscr{C} \rangle$ denote the set of isomorphism classes of objects of $\mathscr{C}$. By abuse of language by an exact sequence in $\mathscr{C}$ we mean an exact sequence $0 \to A' \to A \to A'' \to 0$ in R-mod with A', A, and A'' all in obj $\mathscr{C}$.

Definition 1.1

The *Grothendieck group* of $\mathscr{C}$, denoted by $K_0(\mathscr{C})$ is the abelian group with generators $[A]$ where $A \in \text{obj}\, \mathscr{C}$ and relations $[A] = [A'] + [A'']$ whenever $0 \to A' \to A \to A'' \to 0$ is exact in $\mathscr{C}$.

More formally, let F be the free abelian group with $\langle \text{obj}\, \mathscr{C} \rangle$ as basis and N the subgroup of F generated by $(\langle A \rangle - \langle A' \rangle - \langle A'' \rangle)$ for all exact sequences $0 \to A' \to A \to A'' \to 0$ in $\mathscr{C}$. Then $K_0(\mathscr{C}) = F/N$.

It is clear that $K_0(\mathscr{C})$ has the following universal property. Given any abelian group H and any map $f\colon \langle \text{obj}\,\mathscr{C} \rangle \to H$ satisfying $f(\langle A \rangle) = f(\langle A' \rangle) + f(\langle A'' \rangle)$ whenever $0 \to A' \to A \to A'' \to 0$ is exact in $\mathscr{C}$, then there exists a unique group homomorphism $g\colon K_0(\mathscr{C}) \to H$ satisfying $g([A]) = f(\langle A \rangle)$.

Proposition 1.2

Let A, B be in $\text{obj}\,\mathscr{C}$. Then $[A] = [B]$ in $K_0(\mathscr{C}) \Leftrightarrow$ there exist exact sequences

$$0 \to M' \to M \to M'' \to 0,$$
$$0 \to N' \to N \to N'' \to 0,$$

in $\mathscr{C}$ with $A \oplus N \oplus M' \oplus M'' \simeq B + N' \oplus N'' \oplus M$.

Proof. The implication $\Leftarrow$ is trivial.

$\Rightarrow$: Let F be a free abelian group with $\langle \text{obj}\,\mathscr{C} \rangle$ as basis. Let k, ℓ be integers greater than or equal to 1 and X_i, Y_j for $1 \le i \le k$ and $1 \le j \le \ell$ be objects of $\mathscr{C}$ satisfying $\sum_{i=1}^{k} \langle X_i \rangle = \sum_{j=1}^{\ell} \langle Y_j \rangle$ in F. Then $k = \ell$ and $\langle X_i \rangle = \langle Y_{\sigma(i)} \rangle$ for some permutation σ of $\{1, \ldots, k\}$. In particular, it follows that $\oplus_{i=1}^{k} X_i \simeq \oplus_{j=1}^{\ell} Y_j$.

Now, suppose $[A] = [B]$ in $K_0(\mathscr{C})$. Then $(\langle A \rangle - \langle B \rangle) \in N$, where N is the subgroup of F generated by elements of the form $\langle C \rangle - \langle C' \rangle - \langle C'' \rangle$ where $0 \to C' \to C \to C'' \to 0$ is exact in $\mathscr{C}$. Hence there exist integers $r \ge 0$ and $s \ge 0$ and exact sequences

$$0 \to M_i' \to M_i \to M_i'' \to 0 \qquad \text{for } 1 \le i \le r,$$
$$0 \to N_j' \to N_j \to N_j'' \to 0 \qquad \text{for } 1 \le j \le s$$

(in case $r = 0$, the first family is empty; in case $s = 0$ the second family is empty) in $\mathscr{C}$ with $\langle A \rangle - \langle B \rangle = \sum_{i=1}^{r}(\langle M_i \rangle - \langle M_i' \rangle - \langle M_i'' \rangle) - \sum_{j=1}^{s}(\langle N_j \rangle - \langle N_j' \rangle - \langle N_j'' \rangle)$ in F. Equivalently,

$$\langle A \rangle + \sum_{i=1}^{r} (\langle M_i' \rangle + \langle M_i'' \rangle) + \sum_{j=1}^{s} \langle N_j \rangle$$

$$= \langle B \rangle + \sum_{i=1}^{r} \langle M_i \rangle + \sum_{j=1}^{s} (\langle N_j' \rangle + \langle N_j'' \rangle)$$

in F. It follows from our earlier observation that

$$A \oplus \left(\bigoplus_{i=1}^{r} (M_i' \oplus M_i'') \right) \oplus \left(\bigoplus_{j=1}^{s} N_j \right) \simeq B \oplus \left(\bigoplus_{i=1}^{r} M_i \right) \oplus \left(\bigoplus_{j=1}^{s} (N_j' \oplus N_j'') \right).$$

If $M = \oplus_{j=1}^{r} M_i$, $M' = \oplus_{i=1}^{r} M'$, $M'' = \oplus_{i=1}^{r} M_i''$, and $N = \oplus_{j=1}^{s} N_j$, $N' = \oplus_{j=1}^{s} N_j'$, $N'' = \oplus_{j=1}^{s} N_j''$ then

$$0 \to M' \to M \to M'' \to 0,$$

$$0 \to N' \to N \to N'' \to 0$$

are exact in $\mathscr{C}$ and $A \oplus M' \oplus M'' \oplus N \simeq B \oplus M \oplus N' \oplus N''$. $\square$

Lemma 1.3

Any element of $K_0(\mathscr{C})$ can be written as $[A] - [B]$ for some A, B in obj $\mathscr{C}$.

Proof. It is clear that any element $x \in K_0(\mathscr{C})$ could be written as $x = \sum_{i=1}^{k}[A_i] - \sum_{j=1}^{\ell}[B_j]$, with A_i, B_j in obj $\mathscr{C}$. If $A = \oplus_{i=1}^{k} A_i$ and $B = \oplus_{j=1}^{\ell} B_j$, then $[A] - [B] = x$. $\square$

Definition 1.4

$K_0(R) = K_0(\underline{P}(R))$ *and* $G_0(R) = K_0(\underline{\underline{M}}(R))$.

Let grp($[R]$) denote the subgroup of $K_0(R)$ generated by $[R]$. The quotient group $K_0(R)/\mathrm{grp}([R])$ is denoted by $\tilde{K}_0(R)$. Similarly the quotient of $G_0(R)$ by the subgroup generated by $[R]$ in $G_0(R)$ is denoted by $\tilde{G}_0(R)$. The group $K_0(R)$ [resp. $G_0(R)$] is referred to as the Grothendieck group of finitely generated projective modules (resp. all finitely generated modules) over R. The groups $\tilde{K}_0(R)$ and $\tilde{G}_0(R)$ are referred to as the corresponding reduced Grothendieck groups. Let $\eta: K_0(R) \to \tilde{K}_0(R)$ denote the canonical quotient map. For any $P \in \underline{P}(R)$, by abuse of notation we will denote $\eta([P]) \in \tilde{K}_0(R)$ also by $[P]$; the meaning will be clear from the context. Similar comments apply to $G_0(R)$ and $\tilde{G}_0(R)$.

From Lemma 1.3 we see that an arbitrary element of $K_0(R)$ can be written as $[P] - [Q]$ for suitable P, Q in $\underline{P}(R)$. But $Q \in \underline{P}(R) \Rightarrow \exists$ some $Q' \in \underline{P}(R)$ and an integer $k \geq 1$ with $Q \oplus Q' \simeq R^k$. Then $[P] - [Q] = [P \oplus Q'] - [R^k] = [P'] - k[R]$ where $P' = P \oplus Q'$. Thus we obtain the following:

Corollary 1.5

Any element of $K_0(R)$ can be written as $[A] - k[R]$ for some $A \in \underline{P}(R)$ and integer $k \geq 1$.

Similarly, an element of $K_0(R)$ can be written as $\ell[R] - [B]$ for some integer $\ell \geq 1$ and $B \in \underline{P}(R)$.

As an immediate consequence of Corollary 1.5 we obtain:

Corollary 1.6

Any element of $\tilde{K}_0(R)$ can be written as $[A]$ for a suitable $A \in \underline{P}(R)$.

Again from Lemma 1.3, we see that any element of $G_0(R)$ can be written as $[M] - [N]$ with M, N in $\underline{M}(R)$. Since $N \in \underline{M}(R)$, there exists an exact sequence $0 \to N' \to R^k \to N \to 0$ in R-mod, for some integer $k \geq 1$. In case R is noetherian, we see that $N' \in \underline{M}(R)$, hence $[N'] + [N] = k[R]$ in $G_0(R)$. Thus $[M] - [N] = [M \oplus N'] - k[R]$. Hence we have:

Corollary 1.7

Let R be noetherian. Then any element of $G_0(R)$ can be written as $[A] - k[R]$ for some $M \in \underline{M}(R)$ and integer $k \geq 1$. Any element of $\tilde{G}_0(R)$ can be written as $[A]$ for a suitable $A \in \underline{M}(R)$.

Let P, Q be $\underline{P}(R)$. From Proposition 1.2 we see that $[P] = [Q]$ in $K_0(R)$ if and only if there exist exact sequences

$$0 \to M' \to M \to M'' \to 0,$$
$$0 \to N' \to N \to N'' \to 0,$$

in $\underline{P}(R)$ with $P \oplus M' \oplus M'' \oplus N \simeq Q \oplus M \oplus N' \oplus N''$. Since M'' and N'' are in $\underline{P}(R)$, the two preceding sequences split, yielding $M \simeq M' \oplus M''$ and $N \simeq N' \oplus N''$. Let $A = M' \oplus M'' \oplus N' \oplus N''$. Then $A \in \underline{P}(R)$ and $P \oplus A \simeq Q \oplus A$. Moreover, $\exists$ some $B \in \underline{P}(R)$ and an integer $k \geq 1$ with $A \oplus B \simeq R^k$. Hence $P \oplus R^k \simeq P \oplus A \oplus B \simeq Q \oplus A \oplus B \simeq Q \oplus R^k$. Thus we get the following:

Proposition 1.8

Let P, Q be in $\underline{P}(R)$. Then $[P] = [Q]$ in $K_0(R) \Leftrightarrow \exists$ some integer $k \geq 1$ with $P \oplus R^k \simeq Q \oplus R^k$.

As an immediate corollary of Proposition 1.8 we get:

Corollary 1.9

Let P, Q be in $\underline{P}(R)$. Then $[P] = [Q]$ in $\tilde{K}_0(R) \Leftrightarrow \exists$ integers $k \geq 1$ and $\ell \geq 1$, with $P \oplus R^k \simeq Q \oplus R^\ell$.

Definition 1.10

Two R-modules M and N are said to be stably isomorphic if there exists an integer $k \geq 1$ with $M \oplus R^k \simeq N \oplus R^k$.

From Corollary 1.9 we see that $[P] = 0$ in $\tilde{K}_0(R) \Leftrightarrow P \oplus R^k \simeq R^\ell$ for suitable integers $k \geq 1$ and $\ell \geq 1$. Any such $P \in \underline{P}(R)$ is said to be stably free.

2　INDUCED HOMOMORPHISMS AND RESTRICTION MAPS

Let $f: R \to S$ be a homomorphism of rings. Using f, we can regard S as a right R-module. Namely, $sr = sf(r)$ for any $s \in S$ and $r \in R$. For any $M \in R$-mod, $S \otimes_R M$ is in S-mod. It is clear that $M \in \underline{M}(R) \Rightarrow S \otimes_R M \in \underline{M}(S)$. Also from $S \otimes_R R \simeq S$ in S-mod and the fact that $S \otimes_R$ –commutes with direct sums we immediately get the implication $P \in \underline{P}(R) \Rightarrow S \otimes_R P \in \underline{P}(S)$. Since every short exact sequence in $\underline{P}(R)$ splits, we immediately see that the exactness of $0 \to P' \to P \to P'' \to 0$ in $\underline{P}(R)$ implies the exactness of $0 \to S \otimes_R P' \to S \otimes_R P \to S \otimes_R P'' \to 0$ in $\underline{P}(S)$. Thus $[P] |\rightsquigarrow [S \otimes_R P]$ gives rise to a homomorphism $K_0(R) \to K_0(S)$, which we denote by f_*. Also, since $S \otimes_R R \simeq S$ in S-mod, f_* passes down to quotients to yield a homomorphism $\tilde{K}_0(R) \to \tilde{K}_0(S)$, which will also be denoted by f_*.

If $0 \to M' \to M \to M'' \to 0$ is exact in $\underline{M}(R)$, in general it is *not* true that $0 \to S \otimes_R M' \to S \otimes_R M \to S \otimes_R M'' \to 0$ is exact in S-mod. Recall that S is flat as a right R-module if and only if $S \otimes_R -: R$-mod $\to S$-mod is an exact functor. Thus when S is flat as a right R-module, $[M] |\rightsquigarrow [S \otimes_R M]$ gives rise to a homomorphism $G_0(R) \to G_0(S)$, which we denote by f_*. Since $S \otimes_R R \simeq S$, we get an induced homomorphism $\tilde{G}_0(R) \to \tilde{G}_0(S)$ also denoted by f_*.

For any $P \in \underline{P}(R)$, the map carrying the class $[P]$ of P in $K_0(R)$ to the class of P in $G_0(R)$ [by abuse of notation we will denote the class of P in $G_0(R)$ by $[P] \in G_0(R)$] yields a well-defined homomorphism $c: K_0(R) \to G_0(R)$ called the Cartan homomorphism. Clearly c induces a map $\tilde{K}_0(R) \to \tilde{G}_0(R)$, which will also be denoted by c. When S is flat in mod-R, then

$$\begin{array}{ccc} K_0(R) & \xrightarrow{f_*} & K_0(S) \\ c\downarrow & & \downarrow c \\ G_0(R) & \xrightarrow{f_*} & G_0(S) \end{array} \quad \text{and} \quad \begin{array}{ccc} \tilde{K}_0(R) & \xrightarrow{f_*} & \tilde{K}_0(S) \\ c\downarrow & & \downarrow c \\ \tilde{G}_0(R) & \xrightarrow{f_*} & \tilde{G}_0(S) \end{array}$$

are clearly commutative.

Using $f: R \to S$ we can regard any $N \in S$-mod as an R-module, namely $rx = f(r)x$ for all $r \in R$ and $x \in N$. In particular S itself acquires the structure of a left R-module. Suppose $S \in \underline{M}(R)$. Then any $N \in \underline{M}(S)$ (regarded as a left-module using f) is automatically in $\underline{M}(R)$. If $\varphi: N' \to N$ is any map in S-mod, then automatically φ is a map in R-mod also. In particular $0 \to N' \xrightarrow{\mu} N \xrightarrow{\varepsilon} N'' \to 0$ exact in S-mod $\Rightarrow 0 \to N' \xrightarrow{\mu} N \xrightarrow{\varepsilon} N'' \to$

0 exact in R-mod. Hence we get a well-defined homomorphism $f^*: G_0(S) \to G_0(R)$ satisfying $f^*([N]) = [N] \in G_0(R)$ for any $N \in \underline{M}(S)$. This map f^* is said to be obtained by restriction of scalars.

Suppose $S \in \underline{P}(R)$. Then any $A \in \underline{P}(S)$ (regarded as a left R-module using f) is automatically in $\underline{P}(R)$. Thus we get a well-defined homomorphism $f^*: K_0(S) \to K_0(R)$ satisfying $f^*([A]) = [A] \in K_0(R)$ for any $A \in \underline{P}(S)$. Moreover the diagram

$$
\begin{array}{ccc}
K_0(S) & \xrightarrow{f^*} & K_0(R) \\
c \downarrow & & \downarrow c \\
G_0(S) & \xrightarrow{f^*} & G_0(R)
\end{array}
$$

is clearly commutative.

Now, suppose S is stably free in R-mod. Then the maps $f^*: G_0(S) \to G_0(R)$ and $f^*: K_0(S) \to K_0(R)$ induce maps $f^*: \tilde{G}_0(S) \to \tilde{G}_0(R)$, and $f^*: \tilde{K}_0(S) \to \tilde{K}_0(R)$. The diagram

$$
\begin{array}{ccc}
\tilde{K}_0(S) & \xrightarrow{f^*} & \tilde{K}_0(R) \\
c \downarrow & & \downarrow c \\
\tilde{G}_0(S) & \xrightarrow{f^*} & \tilde{G}_0(R)
\end{array}
$$

is clearly commutative.

3   SOME EXAMPLES

Example 3.1

Let Z denote the ring of integers. Then any $P \in \underline{P}(Z)$ is isomorphic to $Z^n = Z \oplus \cdots \oplus Z$ (direct sum of n copies of Z) where n is an integer greater than or equal to 0, completely determined by $\langle P \rangle$, called the rank of P. Writing rk P for rank P, we have rk $P = $ rk $P' + $ rk P'' whenever $0 \to P' \to P \to P'' \to 0$ is exact in $\underline{P}(Z)$. From the universal property satisfied by K_0, there exists a unique homomorphism $r: K_0(Z) \to Z$ satisfying $r([P]) = $ rk P for any $P \in \underline{P}(R)$. From P isomorphic to the direct sum of rk P copies of Z we infer that $[P] = $ rk $P[Z]$ in $K_0(Z)$. The map $\beta: Z \to K_0(Z)$ given by $\beta(m) = m[Z] \in K_0(Z)$ for any $m \in Z$ clearly satisfies $r\beta = \text{Id}_Z$, $\beta r = \text{Id}_{K_0(Z)}$. Hence $r: K_0(Z) \simeq Z$.

It follows immediately that $\tilde{K}_0(Z) = 0$.

Example 3.2

Let Z denote the ring of integers. Let $A \in \underline{M}(Z)$. From the structure theorem for finitely generated abelian groups, we have $A \simeq Z^n \oplus Z/d_1Z$

$\oplus \cdots \oplus Z/d_k Z$ where n is an integer greater than or equal to 0 and $d_1, \ldots, d_k$ are integers greater than or equal to 1 satisfying $d_1/d_2/d_3/ \cdots /d_k$. (k can very well be 0 in which case $A \simeq Z^n$.) Moreover n, $d_1, \ldots, d_k$, are completely determined by $\langle A \rangle$. n is called the rank of A and is denoted by rk A. In fact rk A equals the dimension of the vector space $Q \otimes_Z A$ over Q where Q is the field of rational numbers.

When $A \simeq Z^n \otimes Z/d_1 Z \oplus \cdots \oplus Z/d_k Z$, we have $[A] = n[Z] + \sum_{j=1}^{k}[Z/d_j Z]$ in $G_0(Z)$. Also the exactness of $0 \to Z \xrightarrow{d_j} Z \to Z/d_j Z \to 0$ in $\underline{M}(Z)$ implies that $[Z] + [Z/d_j Z] = [Z]$ in $G_0(Z)$. Hence $[Z/d_j Z] = 0$. This yields $[A] = n[Z]$ in $G_0(Z)$. Also if $0 \to A' \to A \to A'' \to 0$ is exact in $\underline{M}(Z)$, then rk A = rk A' + rk A''. Hence there exists a unique homomorphism $r: G_0(Z) \to Z$ satisfying $r([A]) =$ rk A for any $A \in \underline{M}(Z)$.

If $\beta: Z \to G_0(Z)$ is defined by $\beta(m) = m[Z]$ in $G_0(Z)$, then it is clear that $r \circ \beta = \mathrm{Id}_Z$ and $\beta \circ r = \mathrm{Id}_{G_0(Z)}$. Hence $r: G_0(Z) \simeq Z$.

Also from Examples 3.1 and 3.2 we see immediately that $c: K_0(Z) \to G_0(Z)$ is an isomorphism.

Examples 3.1 and 3.2 can be generalized to any principal ideal domain (PID).

Example 3.3

Let V be an infinite-dimensional vector space over a field K and $R = \mathrm{End}_K V = \mathrm{Hom}_K(V, V)$. We will show that $K_0(R) = 0$.

Choose a K isomorphism $\varphi: V \to V \oplus V$. Then $\varphi^*: \mathrm{Hom}_K(V \oplus V, V) \to \mathrm{Hom}_K(V, V)$ is a K isomorphism. We can identify $\mathrm{Hom}_K(V \oplus V, V)$ with $\mathrm{Hom}_K(V, V) \oplus \mathrm{Hom}_K(V, V)$. More explicitly let $p_1: V \oplus V \to V$ and $p_2: V \oplus V \to V$ be the projections onto the first and second factors, respectively. Then $\varphi^*: R \oplus R \to$ is given by

$$\{\varphi^*(f, g)\}(v) = f(p_1\varphi(v)) + g(p_2\varphi(v))$$

for any f, g in R and any $v \in V$. We know that φ^* is a K isomorphism. We claim that it is actually an isomorphism of R-modules. In fact, for any $h \in R$ we have

$$\{\varphi^*(hf, hg)\}(v) = hf(p_1\varphi(v)) + hg(p_2\varphi(v))$$
$$= h(f(p_1\varphi(v))) + g(p_2\varphi(v))$$
$$= \{h\varphi^*(f, g)\}(v).$$

Hence $\varphi^*(hf, hg) = h\varphi^*(f, g)$. This proves that $\varphi^*: R \oplus R \simeq R$ in R-mod. It follows by induction on n that $R^n \simeq R$ in R-mod for all integers $n \geq 1$.

From $R \oplus R \simeq R$ in R-mod, we get $[R] + [R] = [R]$ in $K_0(R)$. Hence $[R] = 0$ in $K_0(R)$.

Now let $P \in \underline{P}(R)$. Then there exists a P' in $\underline{P}(R)$ with $P \oplus P' \simeq R^n$ for a suitable n. Since $R^n \simeq R$ we get $P \oplus P' \simeq R$. Thus replacing P, P' by isomorphic copies we may assume that P, P' are left ideals in R with $P \oplus P' = R$. This means $P = Re$, $P' = R(1 - e)$, where e is an idempotent in R. Thus $e: V \to V$ is an endomorphism with $e^2 = e$. It follows that $V = e(V) \oplus (1 - e)(V)$. Observe that $\mathrm{Hom}_K(e(V), V)$ and $\mathrm{Hom}_K((1 - e)(V), V)$ are R-modules. For any $\alpha: e(V) \to V$ and $f: V \to V$ we have $f\alpha: e(V) \to V$. The decomposition $V = e(V) \oplus (1 - e)(v)$ yields an R-isomorphism $\mathrm{Hom}_K(V, V) = R$ with $\mathrm{Hom}_K(e(V), V) \oplus \mathrm{Hom}_K((1 - e)(V), V)$. Under this isomorphism, $\mathrm{Hom}_K(e(V), V)$ corresponds to Re and $\mathrm{Hom}_K((1 - e)(V), V)$ corresponds to $R(1 - e)$, as can easily be checked. Since V is infinite-dimensional either $e(V) \simeq V$ or $(1 - e)(V) \simeq V$ as vector spaces over K. If $e(V) \simeq V$ we get $Re \simeq R$. Hence $[P] = [R] = 0$ in $K_0(R)$. If $(1 - e)(V) \simeq V$, we get $[P'] = [R] = 0$. But then, $[P] + [P'] = [R] = 0 \Rightarrow [P] = 0$. Thus $K_0(R) = 0$.

Example 3.4

Suppose R is a commutative ring. Then any left R-module can be regarded in a natural way as a right R-module and vice versa. If $M \in R$-mod, then the right action of R on M is given by $x \cdot r = rx$ for any $x \in M$ and $r \in R$. Thus we will not distinguish between left and right R-modules in this case. Also for any two R-modules M, N there is a canonical R-module structure on $M \otimes_R N$, given by $r(x \otimes_R y) = (rx \otimes_R y) = (x \otimes_R ry)$. Also, for any three R-modules A, B, C we have $(A \otimes_R B) \otimes_R C \simeq A \otimes_R (B \otimes_R C)$ as R-modules.

Suppose $P \in \underline{P}(R)$, $Q \in \underline{P}(R)$. Then there exist $P' \in \underline{P}(R)$, $Q' \in \underline{P}(R)$, and integers $k \geq 1$ and $\ell \geq 1$, with $P \oplus P' \simeq R^k$ and $Q \oplus Q' \simeq R^\ell$. It follows that $P \otimes_R Q$ is a direct summand of $R^k \otimes_R R^\ell \simeq R^{k\ell}$. Thus $P \otimes_R Q \in \underline{P}(R)$. We can convert $K_0(R)$ into a ring by setting $[P] \cdot [Q] = [P \otimes_R Q]$. More formally, the free abelian group F on isomorphism classes of f.g. (finitely generated) projective R-modules acquires the structure of a ring if we set $\langle P \rangle \cdot \langle Q \rangle = \langle P \otimes_R Q \rangle$. The subgroup N of F generated by $\langle P \oplus Q \rangle - \langle P \rangle - \langle Q \rangle$ with P, Q in $\underline{P}(R)$ is an ideal of F. Hence $K_0(R) = F/N$ is a ring. It is a commutative ring since $P \otimes_R Q \simeq Q \otimes_R P$ in R-mod. The class $[R]$ of R is the identity element of $K_0(R)$.

Let R, S be commutative rings and $f: R \to S$ be a ring homomorphism. For any M, N in R-mod, we have

$$(S \otimes_R M) \otimes_S (S \otimes_R N) \simeq (M \otimes_R S) \otimes_S (S \otimes_R N)$$
$$\simeq M \otimes_R (S \otimes_S S) \otimes_R N$$
$$\simeq (M \otimes_R S) \otimes_R N$$
$$\simeq S \otimes_R (M \otimes_R N).$$

It follows that $K_0(R) \xrightarrow{f_*} K_0(S)$ is a ring homomorphism.

4   FUNCTORIALITY OF K_0

Let $f: R \to S$ and $g: S \to T$ be ring homomorphisms. We regard S as a right R-module using f, T as a right S-module using g, and T as a right R-module using $g \circ f: R \to T$. For any $A \in R$-mod, we have

$$T \otimes_S (S \otimes_R A) \simeq (T \otimes_S S) \otimes_R A \simeq T \otimes_R A.$$

From this it follows that $(g \circ f)_* = g_* \circ f_*: K_0(R) \to K_0(T)$. From $R \otimes_R A \simeq A$ for any A in R-mod, we see that $\mathrm{Id}_*: K_0(R) \to K_0(R)$ is $\mathrm{Id}_{K_0(R)}$. Thus K_0 is a covariant functor from the category of rings to the category of abelian groups. Also from (3.4) we see that, on the category of commutative rings, K_0 is a functor into commutative rings.

Lemma 4.1

Let I be any two-sided ideal of R and $\eta: R \to R/I$ the canonical quotient map. Then for any $P \in \underline{P}(R)$ we have $\eta_*([P]) = [\overline{P}]$ where $\overline{P} = P/IP$ in R/I-mod.

Proof. The exactness of $0 \to I \overset{i}{\to} R \overset{\eta}{\to} R/I \to 0$ in mod-R implies that $I \otimes_R P \xrightarrow{i \otimes 1_P} R \otimes_T P \xrightarrow{\eta \otimes 1_P} (R/I) \otimes_R P \to 0$ is exact. When we identify $R \otimes_R P$ with P under the isomorphism $r \otimes_R x \mapsto rx$ for $r \in R$, $x \in P$, it is clear that the image of $i \otimes 1_P$ is IP. Hence $(R/I) \otimes_R P \simeq P/IP$. $\square$

Proposition 4.2

For any two rings R, S we have $K_0(R \times S) \simeq K_0(R) \times K_0(S)$.

Proof. Let $\pi_1: R \times S \to R$ and $\pi_2: R \times S \to S$ be the canonical projections. These are ring homomorphisms and as such induce homomorphisms $\pi_{1*}: K_0(R \times S) \to K_0(R)$ and $\pi_{2*}: K_0(R \times S) \to K_0(S)$. For any $P \in \underline{P}(R \times S)$, from Lemma 4.1 we see that $\pi_{1*}([P]) = [P/SP]$ and $\pi_{2*}([P]) = [\overline{P}/RP]$ where we identify S with the ideal $0 \times S$ of $R \times S$ and identify R with the ideal $R \times 0$ of $R \times S$. [*Caution:* $R \times 0$ and $0 \times S$ are not subrings of $R \times S$. The identity element of $R \times 0$ is $(1_R, 0)$ whereas the identity element of $R \times S$ is $(1_R, 1_S)$.]

Given any $A \in S$-mod we regard it as an $R \times S$-mod by $(r, s)a = sa$ $\forall\, a \in A$ and $(r, s) \in R \times S$. Since S regarded as an $R \times S$-module in this way is projective, it follows that any $B \in \underline{P}(S)$ regarded as an $R \times S$-module in the preceding manner is in $\underline{P}(R \times S)$. Write $\tilde{B}$ for this element of $\underline{P}(R \times S)$. If $0 \to B' \to B \to B'' \to 0$ is exact in $\underline{P}(S)$, then clearly $0 \to \tilde{B}' \to \tilde{B} \to \tilde{B}'' \to 0$ is exact in $\underline{P}(R \times S)$. Hence we get a well-defined homomorphism $i_2: K_0(S) \to K_0(R \times S)$ satisfying $i_2([B]) = [\tilde{B}]$. Similarly we have a homomorphism $i_1: K_0(R) \to K_0(R \times S)$.

For any $B \in \underline{P}(S)$, clearly $R\tilde{B} = 0$. Hence $\tilde{B}/R\tilde{B}$ regarded as an S-module is B itself. Thus $\pi_{2*}i_2([B]) = [B]$ for any $B \in \underline{P}(S)$ or $\pi_{2*}i_2 = \mathrm{Id}_{K_0(S)}$. Similarly $\pi_{1*}i_1 = \mathrm{Id}_{K_0(R)}$.

Also for any $B \in \underline{P}(S)$ we have $S\tilde{B} = \tilde{B}$; hence $\tilde{B}|S\tilde{B} = 0$. Thus $\pi_{1*} \circ i_2([B]) = 0$ or $\pi_{1*}i_2 = 0$. Similarly $\pi_{2*}i_1 = 0$.

For any $M \in R \times S$-mod, we have $M = RM \oplus SM$ in $R \times S$-mod. Let $P \in \underline{P}(R \times S)$. Then $\pi_{2*}[P] = [P/RP] \in K_0(S)$. From $P \simeq RP \oplus SP$ we get $\overline{P/RP} \simeq SP$ in $R \times S$-mod. Since $R(P/RP) = 0$, it is clear that the $R \times S$-module P/RP corresponding to the S-module P/RP is the $R \times S$-module $P/RP \simeq SP$. Hence $i_2\pi_{2*}[P] = [SP] \in K_0(R \times S)$. Similarly $i_1\pi_{1*}[P] = [RP] \in K_0(R \times S)$. From $P = RP \oplus SP$ we get $i_1\pi_{1*} + i_2\pi_{2*} = \mathrm{Id}_{K_0}(R \times S)$. This completes the proof of Proposition 4.2. $\square$

Proposition 4.3

Let R be a ring and J denote the Jacobson radical or R. Let $\eta: R \to R/J$ denote the canonical quotient map. Then $\eta_*: K_0(R) \to K_0(R/J)$ is injective.

Proof. Let P, Q be in $\underline{P}(R)$ and $\overline{P} = P/JP, \overline{Q} = Q/JQ$. Let $S = R/J$. Suppose $\overline{P}$ and $\overline{Q}$ are isomorphic as S-modules, say $\theta: \overline{P} \to \overline{Q}$ an S-isomorphism. Regarding $\overline{P}$ and $\overline{Q}$ as R-modules via $\eta: R \to R/J$, it is clear that $\theta: \overline{P} \to \overline{Q}$ is an R-isomorphism.

Let $\pi_1: P \to P/JP$ and $\pi_2: Q \to Q/JQ$ denote the canonical quotient maps. Since P is projective in R-mod and $Q \xrightarrow{\pi_2} Q/JQ \to 0$ is exact in R-mod, we get a map $f: P \to Q$ with

$$
\begin{array}{ccc}
 & & P \\
 & {\scriptstyle f}\swarrow & \downarrow {\scriptstyle \theta \,\circ\, \pi_1} \\
Q & \xrightarrow{\pi_2} & Q/JQ \longrightarrow 0
\end{array}
$$

commutative. We claim that f is an isomorphism. Since π_1 and θ are epimorphisms, so is $\theta \circ \pi_1$. Hence $\pi_2 \circ f: P \to Q/JQ$ is an epimorphism. Hence $f(P) + \ker \pi_2 = Q$ or $f(P) + JQ = Q$. Since Q is f.g. by Nakayama's lemma, $f(P) = Q$. Thus $f: P \to Q$ is an epimorphism. Since Q is projective, there exists a splitting $\mu: Q \to P$ for f, namely $f \circ \mu = \mathrm{Id}_Q$. Thus $P = \mu(Q) \oplus \ker f$. But

$$\ker f \subset \ker \pi_2 \circ f = \ker \theta \circ \pi_1$$

$$= \ker \pi_1 \quad \text{(since } \theta \text{ is an isomorphism)}$$

$$= JP.$$

Hence $\mu(Q) + JP = P$. Since P is f.g., by Nakayama's lemma we get $\mu(Q) = P$. Again, from $P = \mu(Q) \oplus \ker f$ we see that $\ker f = 0$. Hence $f: P \to Q$ is an isomorphism in R-mod.

We now take up the proof that $\eta_*: K_0(R) \to K_0(R/J)$ is injective. Let $x \in K_0(R)$ satisfy $\eta_*(x) = 0$. From Lemma 1.3, we see that $x = [P] - [Q]$ with P, Q in $\underline{P}(R)$. Then $\eta_*(x) = [\bar{P}] - [\bar{Q}] \in K_0(R/J)$ where $\bar{P} = P/JP$ and $\bar{Q} = Q/J\bar{Q}$ by Lemma 4.1. From Proposition 1.8, we see that $\bar{P} \oplus (R/J)^k \simeq \bar{Q} \oplus (R/J)^k$ in R/J-mod for some integer $k \geq 1$. Hence from what is proved already $P \oplus R^k \simeq Q \oplus R^k$. Hence $x = [P] - [Q] = 0$ in $K_0(R)$. $\square$

Corollary 4.4

Let I be a two-sided ideal of R with $I \subset J$ and $\alpha: R \to R/I$ the canonical quotient map. Then $\alpha_*: K_0(R) \to K_0(R/I)$ is injective.

Proof. If $\eta: R \to R/J$ is the canonical quotient map, then η can be factored as $\gamma \circ \alpha$ with $\gamma: R/I \to R/J$ the obvious map. Hence $\eta_* = \gamma_* \circ \alpha_*$. From Proposition 4.3 we have η_* injective. Hence α_* is injective. $\square$

Remark 4.5

Let R be a skew field. Then any $M \in R$-mod is free. Also $P \in \underline{P}(R)$ if and only if P is a finite-dimensional vector space. Since $\dim(P \oplus Q) = \dim P + \dim Q$ we get a well-defined homomorphism $\mathrm{rk}: K_0(R) \to Z$ satisfying $\mathrm{rk}([P]) = \dim P$. It can easily be checked that rk is an isomorphism.

Definition 4.6

R is said to be a local ring if the nonunits in R form a two-sided ideal $\mathfrak{m}$ in R.

In this case $\mathfrak{m}$ is the unique maximal left (or right) ideal of R and thus $J = \mathfrak{m}$. Moreover R/J is a skew field.

Corollary 4.7

If R is a local ring, then $K_0(R) \simeq Z$.

Proof. Let J be the Jacobson radical of R. Then R/J is a skew field; hence $K_0(R/J) \simeq Z$. From Proposition 4.3, we see that $\eta_*: K_0(R) \to K_0(R/J)$ is injective. Also $\eta_*([R]) = [R/J]$ the generator of $K_0(R/J) \simeq Z$. Thus $K_0(R) \simeq Z$ with $[R]$ as the generator. In this case $\eta_*: K_0(R) \to K_0(R/J)$ is an isomorphism. $\square$

Corollary 4.8

Let m be an integer greater than or equal to 2 and let n be the number of distinct primes dividing m. Then $K_0(Z/mZ) \simeq Z^n$ (direct sum of n copies of Z).

Proof. Let $m = p_1^{k_1} \cdots p_n^{k_n}$ with k_i integers greater than or equal to 1 and $p_1, \ldots, p_n$ distinct primes. Then $Z/mZ \simeq Z/p_1^{k_1}Z \times \cdots \times Z/p_n^{k_n}Z$ as rings. Also $Z/p_i^{k_i}Z$ is a local ring. Using Proposition 4.2 and Corollary 4.7 we get

$$K_0(Z/mZ) \simeq K_0\big(Z/p_1^{k_1}Z\big) \times \cdots \times K_0\big(Z/p_n^{k_n}Z\big)$$

$$\simeq Z^n. \qquad \qquad \square$$

5 GROTHENDIECK'S THEOREM

Lemma 5.1 (Schanuel's Lemma)

Let

$$0 \longrightarrow B \xrightarrow{\ i\ } P \xrightarrow{\ \varepsilon\ } A \longrightarrow 0,$$

$$0 \longrightarrow B' \xrightarrow{\ i'\ } P' \xrightarrow{\ \varepsilon'\ } A' \longrightarrow 0,$$

be exact sequences with P, P' projective and $A \simeq A'$. Then $B \oplus P' \simeq B' \oplus P$.

Proof. Suppose $\varphi: A' \to A$ is an isomorphism. If $\varepsilon'' = \varphi \circ \varepsilon'$, then $0 \to B' \xrightarrow{\ i'\ } P' \xrightarrow{\ \varepsilon''\ } A \to 0$ is exact. Hence for proving Schanuel's lemma we may as well assume $A = A'$. Let

$$
\begin{array}{ccc}
H & \xrightarrow{\ \beta\ } & P' \\
{\scriptstyle \alpha}\downarrow & & \downarrow{\scriptstyle \varepsilon'} \\
P & \xrightarrow[\ \varepsilon\ ]{} & A
\end{array}
$$

denote a pull-back diagram, namely

$$H = \{(x, x') \in P \times P' \,|\, \varepsilon(x) = \varepsilon'(x')\}$$

$$\left.\begin{array}{l} \alpha(x, x') = x \\[4pt] \beta(x, x') = x' \end{array}\right\} \quad \text{for any } (x, x') \in H.$$

Given $x \in P$, since ε' is onto A $\exists$ an $x' \in P'$ with $\varepsilon(x) = \varepsilon'(x')$. Then $(x, x') \in H$ and $\alpha(x, x') = x$. Thus $\alpha: H \to P$ is onto. Similarly β is onto.

We have the following commutative diagram with exact rows and columns:

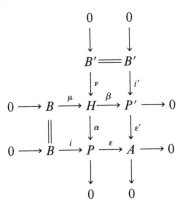

Here

$$\mu(b) = (i(b), 0) \qquad \forall\, b \in b,$$

$$\nu(b') = (0, i'(b')) \qquad \forall\, b' \in B'.$$

Since P is projective, the left vertical exact sequence splits to yield $H \simeq P \oplus B'$. Similarly since P' is projective, the top horizontal exact sequence splits to yield $H \simeq P' \oplus B$.

Hence $P \oplus B' \simeq H \simeq P' \oplus B$. $\square$

Lemma 5.2 (Schamuel's Lemma for Long Exact Sequences)

Let $n \geq 1$ and

$$0 \longrightarrow K_n \longrightarrow P_{n-1} \longrightarrow \cdots \xrightarrow{d_1} P_0 \xrightarrow{\varepsilon} A \longrightarrow 0,$$

$$0 \longrightarrow K_n' \longrightarrow P_{n-1}' \longrightarrow \cdots \xrightarrow{d_1'} P_0' \xrightarrow{\varepsilon'} A' \longrightarrow 0$$

be exact with P_i, P_i' projective for $0 \leq i \leq n - 1$ and $A \simeq A'$.
Then

$$P_0 \oplus P_1' \oplus P_2 \oplus \cdots \oplus X_n \simeq P_0' \oplus P_1 \oplus P_2' \oplus \cdots \oplus Y_n$$

where

$$X_n = \begin{cases} K_n, & \text{if } n \text{ is even,} \\ K_n', & \text{if } n \text{ is odd,} \end{cases}$$

$$Y_n = \begin{cases} K_n, & \text{if } n \text{ is even,} \\ K_n', & \text{if } n \text{ is odd.} \end{cases}$$

Proof. By induction on n.

For $n = 1$, this is Lemma 5.1. Let $n > 1$, $L_0 = \ker \varepsilon = \operatorname{im} d_1$ and $L_0' = \ker \varepsilon' = \operatorname{im} d_1'$. Assume Lemma 5.2 true for $n - 1$ in place of n.

Then $\because$ we have exact sequences

$$0 \longrightarrow K_n \longrightarrow \cdots \longrightarrow P_2 \xrightarrow{(d_2, 0)} P_1 \oplus P_0' \xrightarrow{d_1 \oplus \operatorname{Id}_{P_0'}} L_0 \oplus P_0' \longrightarrow 0,$$

$$0 \longrightarrow K_n' \longrightarrow \cdots \longrightarrow P_2' \xrightarrow{(d_2', 0)} P_1 \oplus P_0 \xrightarrow{d_1' \oplus \operatorname{Id}_{P_0}} L_0' \oplus P_0 \longrightarrow 0,$$

and $L_0 \oplus P_0' \simeq L_0' \oplus P_0$ (because of Lemma 5.1),

hence $P_0' \oplus P_1 \oplus P_2' \oplus \cdots \oplus Y_n \simeq P_0 \oplus P_1' \oplus P_2 \oplus \cdots \oplus X_n.$ $\square$

Corollary 5.3

Suppose $0 \to P_n \to P_{n-1} \to \cdots \to P_0 \to A \to 0$ *and* $0 \to K_n' \to P_{n-1}' \to \cdots \to P_0' \to A \to 0$ *are exact with*

$$P_i \quad \text{projective for } 0 \le i \le n,$$

$$P_j' \quad \text{projective for } 0 \le j \le n - 1,$$

then automatically K_n' *is projective.*

Definition 5.4

Let $A \in R\text{-mod}$. We say that the homological dimension of A over R is less than or equal to m if there exists a "projective resolution" $0 \to P_m \to \cdots \to P_0$ (namely $0 \to P_m \to P_{m-1} \to \cdots \to P_0 \to A \to 0$ is exact and P_i proj for all i) of length m for A.

If $n \ge 0$ is the least integer with the property that A has a projective resolution of length n, we say homological dimension of A over $R = n$ and we write $\operatorname{hd}_R A = n$ or $\operatorname{hd} A = n$ when there is no confusion. If there is no such integer we set $\operatorname{hd} A = \infty$.

Definition 5.5

The left global dimension of R is defined by $\operatorname{l.gl.\,dim} R = \sup_{A \in R\text{-mod}}(\operatorname{hd} A)$.

Remarks 5.6

(i) $\operatorname{hd} A = 0 \Leftrightarrow A$ is projective.

(ii) $\operatorname{l.gl.\,dim} R = 0 \Leftrightarrow R$ is semi-simple artinian.

(iii) $\operatorname{l.gl.\,dim} R \le 1 \Leftrightarrow$ every submodule of a projective R-module is projective. Such rings are known as left hereditary rings.

Lemma 5.7

Let

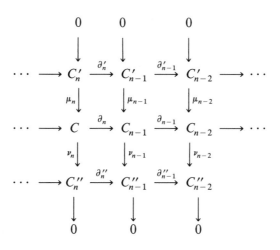

Diagram 5.8.

be a commutative diagram with each vertical sequence exact. Suppose the lower two horizontal sequences are exact. Then the top horizontal sequence is also exact.

Proof. First observe that our assumptions imply that $C'' = \{C_n, \partial_n\}$ and $C'' = \{C_n'', \partial_n''\}$ are chain complexes and that $C = \{C_n', \partial_n'\}$ is a subcomplex of C. Moreover $\mu = (\mu_n)$ and $\nu = (\nu_n)$ yield chain maps $\mu: C' \to C$ and $\nu: C \to C''$, and part of our hypothesis is that

$$0 \longrightarrow C' \overset{\mu}{\longrightarrow} C \overset{\nu}{\longrightarrow} C'' \longrightarrow 0 \qquad (*)$$

is an exact sequence of chain complexes.

Also we are given that $H_n(C) = 0 = H_n(C'')$ for all n. The homology exact sequence of $(*)$ gives us exactness of

$$H_{n+1}(C'') \overset{\partial}{\longrightarrow} H_n(C') \overset{\mu_*}{\longrightarrow} H_n(C) \overset{\nu_*}{\longrightarrow} H_n(C'') \overset{\partial}{\longrightarrow} H_{n-1}(C') \longrightarrow \cdots .$$

It follows that $H_n(C') = 0$ for all n. Hence the top horizontal sequence in Diagram 5.8 is exact. $\square$

Remark 5.9

If each vertical sequence in Diagram 5.8 is exact, from the preceding proof we see that exactness of any two of the horizontal sequences in Diagram 5.8

implies the exactness of the third, provided we assume that C is a chain complex.

Proposition 5.10

Let

$$0 \longrightarrow A' \xrightarrow{j} A \xrightarrow{\eta} A'' \longrightarrow 0,$$

$$0 \longrightarrow P'_k \xrightarrow{d'_k} P'_{k-1} \longrightarrow \cdots \xrightarrow{d'_1} P'_0 \xrightarrow{\varepsilon'} A' \longrightarrow 0,$$

$$0 \longrightarrow P''_k \xrightarrow{d''_k} P''_{k-1} \longrightarrow \cdots \xrightarrow{d''_1} P''_0 \xrightarrow{\varepsilon''} A'' \longrightarrow 0,$$

be exact sequences. Suppose P'_i, P''_i are projective for $0 \le i \le k$. Then $\exists$ an exact sequence of the form

$$0 \longrightarrow P'_k \oplus P''_k \xrightarrow{d_k} P'_{k-1} \oplus P''_{k-1} \longrightarrow \cdots \xrightarrow{d_1} P'_0 \oplus P''_0 \xrightarrow{\varepsilon} A \longrightarrow 0$$

for suitable d_i's and ε.

Proof. We prove this by induction on k if $k = 0$, $P'_0 \simeq A'$, and $P''_0 \simeq A''$. Hence A'' is projective. Hence $A \simeq A' \oplus A''$. Thus the result is valid for $k = 0$. Let $k \ge 1$. From the exactness of $A \xrightarrow{\eta} A'' \to 0$ and projective nature of P''_0, we see that $\exists \; \alpha_0 \colon P''_0 \to A$ satisfying $\eta\alpha_0 = \varepsilon''$. Then the diagram

$$
\begin{array}{ccccccccc}
0 & \longrightarrow & P'_0 & \xrightarrow{i'_0} & P'_0 \oplus P''_0 & \xrightarrow{\pi''_0} & P''_0 & \longrightarrow & 0 \\
& & {\scriptstyle \varepsilon'}\big\downarrow & & {\scriptstyle (j\varepsilon',\,\alpha_0)}\big\downarrow & & \big\downarrow{\scriptstyle \varepsilon''} & & \\
0 & \longrightarrow & A' & \xrightarrow[j]{} & A & \xrightarrow[\eta]{} & A'' & \longrightarrow & 0
\end{array}
$$

Diagram 5.11.

is commutative with exact rows. Here i'_0 is the usual inclusion of P'_0 in $P'_0 \oplus P''_0$ and π''_0 the usual projection of $P'_0 \oplus P''_0$ onto P''_0.

By the five lemma we see that $\varepsilon = (j\varepsilon', \alpha_0) \colon P'_0 \oplus P''_0 \to A$ is onto.

Let

$$B' = \ker \varepsilon' = \operatorname{Im} d'_1,$$

$$B'' = \ker \varepsilon'' = \operatorname{Im} d''_1,$$

$$B = \ker \varepsilon.$$

Let $\mu' \colon B' \to P'_0$, $\mu \colon B \to P'_0 \oplus P''_0$, and $\mu'' \colon B'' \to P''$ be the inclusions. Then $i'_0(B') \subset B$, $\pi''_0(B) \subset B''$ because of commutativity of Diagram 5.11.

Hence we get the following commutative diagram:

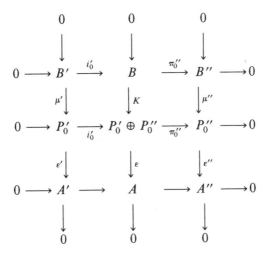

Diagram 5.12.

with vertical sequences exact and the lower two horizontal sequences exact. Hence by Lemma 5.7

$$0 \xrightarrow{\quad} B' \xrightarrow{\;i_0'\;} B \xrightarrow{\;\pi_0''\;} B'' \xrightarrow{\quad} 0 \text{ is exact.}$$

We also know

$$0 \xrightarrow{\quad} P_k' \xrightarrow{\;d_k'\;} P_{k-1}' \xrightarrow{\quad} \cdots \xrightarrow{\quad} P_1' \xrightarrow{\;d_1'\;} B' \xrightarrow{\quad} 0,$$

$$0 \xrightarrow{\quad} P_k'' \xrightarrow{\;d_k''\;} P_{k-1}'' \xrightarrow{\quad} \cdots \xrightarrow{\quad} P_1'' \xrightarrow{\;d_1''\;} B'' \xrightarrow{\quad} 0$$

are exact. By the inductive assumption $\exists$ an exact sequence of the form

$$0 \to P_k' \oplus P_k'' \to \cdots \to P_1' \oplus P_1'' \to B \to 0.$$

Since $B = \ker \varepsilon$ we get the exact sequence

$$0 \xrightarrow{\quad} P_k' \oplus P_k'' \xrightarrow{\quad} \cdots \xrightarrow{\quad} P_1' \oplus P_1'' \xrightarrow{\quad} P_0' \oplus P_0'' \xrightarrow{\;\varepsilon\;} A \xrightarrow{\quad} 0.$$

This proves Proposition 5.10. $\square$

Corollary 5.13

Let $0 \to A' \to A \to A'' \to 0$ be exact with hd $A' = n' < \infty$ and hd $A'' = n < \infty$. Then hd $A \le \max(n', n)$.

Definition 5.14

$A \in R$-mod will be called an FP-module if $\exists$ an exact sequence

$$0 \to P_n \to P_{n-1} \to \cdots \to A \to 0$$

with $n < \infty$ and P_i *finitely generated* projective.

Corollary 5.15

Let $0 \to A' \to A \to A'' \to 0$ *be exact with* A' *and* A'' *modules of type* FP. *Then* A *is also an* FP-*module*.

Let $\underline{P}(R)$ and $\underline{M}(R)$ have their usual meanings. They are full, additive subcategories of R-mod. Let $\underline{H}(R)$ denote the full subcategory of R-mod consisting of modules of type $\overline{\text{FP}}$. Then $\underline{H}(R)$ is an additive subcategory of R-mod, by Corollary 5.15. Thus we can talk of the Grothendieck group $K_0(\underline{H}(R))$ of $\underline{H}(R)$. We have an inclusion $\underline{P}(R) \xrightarrow{j} \underline{H}(R)$ that induces a map $K_0(R) \xrightarrow{j_*} K_0(\underline{H}(R))$.

Theorem 5.16 (Grothendieck)

The map $j_*: K_0(R) \to K_0(\underline{H}(R))$ *is an isomorphism*.

Proof. Let $M \in \underline{H}(R)$. Let

$$0 \to P_n \to \cdots \to P_0 \to M \to 0$$

be exact with $p_i \in \underline{P}(R)$ and $n < \infty$. Lemma 5.2 implies that $\Sigma(-1)^i[P_i] \in K_0(R)$ depends only on M.

Let us define $f: K_0(\underline{H}(R)) \to K_0(\underline{P}(R)) = K_0(R)$ by $f([M]) = \Sigma(-1)^i[P_i] \in K_0(R)$. To see that f is a well-defined homomorphism of abelian groups we have to check that if

$$0 \to M' \to M \to M'' \to 0$$

is any exact sequence in $\underline{H}(R)$ [namely M', M, M'' in $\underline{H}(R)$], then $f([M]) = f([M']) + f([M''])$. $\exists$ an $n < \infty$ and exact sequences

$$0 \to P'_n \to \cdots \to P'_0 \to M' \to 0,$$

$$0 \to P''_n \to \cdots \to P''_0 \to M'' \to 0,$$

with P'_i, P''_i in $\underline{P}(R)$ for $0 \le i \le n$.

By Proposition 5.10 $\exists$ an exact sequence

$$0 \to P_n' \oplus P_n'' \to \cdots \to P_0' \oplus P_0'' \to M \to 0.$$

Hence

$$f([M]) = \sum(-1)^i[P_i' \oplus P_i''] \in K_0(R)$$

$$= \sum(-1)^i[P_i'] + \sum(-1)^i[P_i''] = f([M']) + f([M'']).$$

Thus f is a well-defined homomorphism of abelian groups.

If $P \in \underline{P}(R)$, then $0 \to P_0 \to P \to 0$ is exact where $P_0 = P$. Thus $fj_*([P])$ $= [P] \in \overline{K}_0(R)$. In other words $fj_* = \mathrm{Id}_{K_0(R)}$.

Let $N \in \underline{H}(R)$ and

$$0 \longrightarrow P_n \longrightarrow P_{n-1} \longrightarrow \cdots \xrightarrow{d_1} P_0 \xrightarrow{\varepsilon} N \longrightarrow 0$$

be exact with $P_i \in \underline{P}(R)$ ($0 \le i \le n$). Then each $P_i \in \underline{H}(R)$ also. We will prove by induction on n that $\sum_{i=0}^n(-1)^i[P_i]$ of $K_0(\underline{H}(R))$ is the same as $[N]$. If $n = 0$, $P_0 \simeq N$ and $[P_0] = [N]$ in $K_0(\underline{H}(R))$. Let $n \ge 1$ and assume the result valid for $n - 1$ in place of n. Then $B_0 = \ker \varepsilon \colon P_0 \to N = \mathrm{Im}\, P_1 \xrightarrow{d_1} P_0$. Then $0 \to P_n \to \cdots \to P_1 \xrightarrow{d_1} B_0 \to 0$ is exact. Hence $B_0 \in \underline{H}(R)$. By inductive assumption $[B_0] = \sum_{i=1}^n(-1)^{i-1}[P_i]$ in $K_0(\underline{H}(R))$. Also $0 \to B_0 \to P_0$ $\to N \to 0$ is exact in $\underline{H}(R)$. Hence $[P_0] = [B_0] + [N]$ in $K_0(\underline{H}(R))$. Thus $[P_0] - [B_0] = [N]$ or $[N] = \sum_{i=0}^n(-1)^i[P_i]$ in $K_0(\underline{H}(R))$. This proves $j_* \circ f = \mathrm{Id}_{K_0(\underline{H}(R))}$. $\square$

Definition 5.17

A ring R is called left regular if R is left noetherian and every $M \in \underline{M}(R)$ is of finite homological dimension.

Let R be left regular and $M \in \underline{M}(R)$. Let $n = \mathrm{hd}\, M$. Since R is noetherian every finitely generated module is finitely presented. Using this fact we can get an exact sequence

$$0 \to K_n \to P_{n-1} \to \cdots \to P_0 \to M \to 0$$

with $P_i \in \underline{P}(R)$ for $0 \le i \le n - 1$. Corollary 5.3 yields K_n projective. But K_n is finitely generated. Hence $K_n \in \underline{P}(R)$. This proves that when R is left regular, $\underline{M}(R) = \underline{H}(R)$. From Theorem 5.16 we get the following corollary:

Corollary 5.18

Let R be a left regular ring. Then the Cartan map $c: K_0(R) \to G_0(R)$ is an isomorphism. Since c carries $[R]$ of $K_0(R)$ to $[R]$ of $G_0(R)$, it follows that $c: \tilde{K}_0(R) \to \tilde{G}_0(R)$ is also an isomorphism.

Any noetherian ring of finite global dimension is an example of a regular ring. In particular noetherian hereditary rings are regular.

The property of a left regular ring used in Corollary 5.18 is the equality $\underline{\underline{M}}(R) = \underline{\underline{H}}(R)$. In fact Corollary 5.18 is valid for every ring R for which the equality $\overline{\underline{M}}(R) = \underline{\underline{H}}(R)$ is valid. However, as we will show presently, the equality $\underline{\underline{M}}(R) = \underline{\underline{H}}(R)$ implies R regular. Actually we will be proving the following theorem for any ring R, and from that deducing that $\underline{\underline{M}}(R) = \underline{\underline{H}}(R) \Rightarrow R$ regular.

Theorem 5.19

Let $0 \to M' \overset{j}{\to} M \overset{\eta}{\to} M'' \to 0$ be exact. If any two of the modules M', M, M'' are FP then the third one also is.

The proof of this theorem will involve the construction of the mapping cone of two chain complexes.

Definition 5.20

Let $C = \{C_n, \partial_n\}$ be any chain complex over Λ. We define the suspension ΣC of C to be the chain complex where $(\Sigma C)_n = C_{n-1}$ and $(\Sigma \partial)_n = -\partial_{n-1}: C_{n-1} \to C_{n-2}$.

Definition 5.21

Let $f = (f_n)$ be a chain map of $C = \{C_n, \partial_n\}$ into $C' = \{C'_n, \partial'_n\}$. The "mapping cone" of f, denoted by $M(f)$ is the chain complex given by $M(f)_n = C'_n \oplus C_{n-1}$ where $d_n: M(f)_n \to M(f)_{n-1}$ is $d_n(u', x) = (\partial'_n u' + f_{n-1}(x), -\partial_{n-1}x)$ for any $(u', x) \in C'_n \oplus C_{n-1}$.

In fact

$$d_{n-1}d_n(u', x) = \left(\partial'_{n-1}(\partial' u' + f_{n-1}(x)) + f_{n-2}(-\partial_{n-1}x), \partial_{n-2}\partial_{n-1}x\right)$$

$$= \left(\partial'_{n-1}f_{n-1}(x) - f_{n-2}(\partial_{n-1}x), 0\right)$$

$$= (0, 0).$$

Thus $M(f) = \{M(f)_n, d_n\}$ is a chain complex.

Let $\mu_n\colon C_n' \to M(f)_n$ and $\eta_n\colon M(f)_n \to (\Sigma C) = C_{n-1}$ be defined by

$$\mu_n(u') = (u',0) \in C_n' \oplus C_{n-1},$$

$$\eta_n(u',x) = x \in (\Sigma C)_n = C_{n-1}.$$

Then

$$d_n \mu_n(u') = d_n(u',0) = (\partial_n' u',0) = \mu_{n-1}(\partial_n' u') \tag{5.22}$$

and

$$(\Sigma\partial)_n \circ \eta_n(u',x) = -\partial_{n-1}x = \eta_{n-1}d_n(u',x). \tag{5.23}$$

(5.22) and (5.23) imply that

$$\mu = (\mu_n)\colon C' \to M(f),$$

$$\eta = (\eta_n)\colon M(f) \to \Sigma C$$

are chain maps. Also it is clear that

$$0 \longrightarrow C' \xrightarrow{\ \mu\ } M(f) \xrightarrow{\ \eta\ } \Sigma C \longrightarrow 0 \tag{$**$}$$

is an exact sequence of chain complexes. It is also clear that $H_n(\Sigma C) = H_{n-1}(C)$ [in fact $Z_n(\Sigma C) = Z_{n-1}(C)$ and $B_n(\Sigma C) = B_{n-1}(C)$].

Lemma 5.24

The boundary map

$$\partial_*\colon H_n(\Sigma C) \longrightarrow H_{n-1}(C')$$
$$\|$$
$$H_{n-1}(C)$$

in the homology exact sequence of $(**)$ *is the same as* $f_*\colon H_{n-1}(C) \to H_{n-1}(C')$ *induced by* $f\colon C \to C'$.

Proof. Let $a \in H_n(\Sigma C) = H_{n-1}(C)$. Let $z \in Z_n(\Sigma C) = Z_{n-1}(C)$ represent a. Then $(0,z) \in M(f)_n = C_n' \oplus C_{n-1}$ satisfies $\eta_n(0,z) = z$. Now $d_n(0,z) = (f_{n-1}(z),0)$ and hence $\mu_{n-1}(f_{n-1}(z)) = d_n(0,z)$. By definition $\partial_* a = [f_{n-1}(z)]$ the homology class of $f_{n-1}(z)$ in $H_{n-1}(C)$. But $f_*(a)$ equals the homology class of $f_{n-1}(z)$ in $H_{n-1}(C)$. Thus $\partial_* = f_*$. $\square$

Definition 5.25

A positive chain complex $C = \{C_n, \partial_n\}_{n \geq 0}$ provided with an augmentation $\varepsilon: C_0 \to N$ will be called a resolution of N if

$$\cdots \longrightarrow C_n \longrightarrow C_{n-1} \longrightarrow \cdots \longrightarrow C_1 \longrightarrow C_0 \xrightarrow{\varepsilon} N \longrightarrow 0$$

is exact. In this situation we will just write $C \xrightarrow{\varepsilon} N$ is a resolution of N.

Let $C \xrightarrow{\varepsilon} N$ and $C' \xrightarrow{\varepsilon'} N'$ be resolutions of N and N' and $f: N \to N'$ a map.

Definition 5.26

A chain map $\varphi: C \to C'$ is said to be over f (or said to cover f) if the following diagram is commutative:

$$
\begin{array}{ccccccccccc}
\cdots & \longrightarrow & C_n & \longrightarrow & C_{n-1} & \longrightarrow & \cdots & \longrightarrow & C_1 & \xrightarrow{\partial_1} & C_0 & \xrightarrow{\varepsilon} & N & \longrightarrow & 0 \\
& & \downarrow{\scriptstyle\varphi_n} & & \downarrow{\scriptstyle\varphi_{n-1}} & & & & \downarrow{\scriptstyle\varphi_1} & & \downarrow{\scriptstyle\varphi_0} & & \downarrow{\scriptstyle f} & & \\
\cdots & \longrightarrow & C_n' & \longrightarrow & C_{n-1}' & \longrightarrow & \cdots & \longrightarrow & C_1' & \xrightarrow{\partial_1'} & C_0' & \xrightarrow{\varepsilon'} & N' & \longrightarrow & 0
\end{array}
$$

Diagram 5.27.

Theorem 5.28

Let $C \xrightarrow{\varepsilon} N$ and $C' \xrightarrow{\varepsilon'} N'$ be resolutions, $f: N \to N'$ a map, and $\varphi: C \to C'$ a chain map covering f. Let $\alpha: N' \to N'/f(N)$ denote the canonical quotient map. Let $\bar{\varepsilon} = \alpha \circ \varepsilon': C_0' = M(\varphi)_0 \to N'/(f(N))$.

(i) *If f is a monomorphism, $M(\varphi) \xrightarrow{\bar{\varepsilon}} N'/f(N)$ is a resolution of $N'/f(N)$.*

(ii) *If f is an epimorphism, there is a resolution of $\ker f$ of the form*

$$\cdots \to M(\varphi)_n \to \cdots \to M(\varphi)_3 \to M(\varphi)_2 \to Z_1(M(\varphi)) \to \ker f \to 0.$$

Proof. Consider the homology exact sequence of $0 \to C' \xrightarrow{\mu} M(\varphi) \xrightarrow{\eta} \Sigma C \to 0$, namely the exact sequence

$$\cdots \longrightarrow H_n(C') \xrightarrow{u_*} H_n(M(\varphi)) \xrightarrow{\eta_*} H_{n-1}(C) \xrightarrow{\varphi_*} H_{n-1}(C') \longrightarrow \cdots$$

$$\longrightarrow H_1(C') \xrightarrow{\mu_*} H_1(M(\varphi)) \xrightarrow{\eta_*} H_0(C) \longrightarrow H_0(C') \longrightarrow H_0(M(\varphi)) \longrightarrow 0$$

$$
\begin{array}{ccc}
\| & & \| \\
N & \xrightarrow{f} & N'
\end{array}
$$

We know that $H_i(C) = 0 = H_i(C')$ for $i \geq 1$. Hence $H_j(M(\varphi)) = 0$ for $j \geq 2$. It follows also, that if f is a monomorphism, then $H_1(M(\varphi)) = 0$ and $H_0(M(\varphi)) \simeq N'/f(N)$. This actually proves (i).

If f is an epimorphism, $H_0(M(\varphi)) = 0$, $H_1(M(\varphi)) \simeq \ker f$, and $H_j(M(\varphi)) = 0$ for $j \geq 2$. Hence we get exact sequences

$$0 \to Z_1(M(\varphi)) \to M(\varphi)_1 \to M(\varphi)_0 \to 0,$$

$$\cdots \to M(\varphi)_n \cdots \to M(\varphi)_3 \to M(\varphi)_2 \to Z_1(M(\varphi)) \to \ker f \to 0.$$

This proves (ii). $\square$

Proof of Theorem 5.19

(i) We have already seen that if M', M'' are FP, then so is M (Corollary 5.15).

(ii) Now assume M and M'' are FP. Let

$$
\begin{array}{ccccccccc}
0 & \longrightarrow & P_n'' & \longrightarrow & \cdots & \xrightarrow{\partial_1''} & P_0'' & \xrightarrow{\varepsilon''} & M'' & \longrightarrow & 0 \\
& & & & & & & & \uparrow{\eta} & & \\
0 & \longrightarrow & P_k & \longrightarrow & \cdots & \xrightarrow{\partial_1} & P_0 & \xrightarrow{\varepsilon} & M & \longrightarrow & 0
\end{array}
$$

be resolutions of finite length with P_i'', P_j in $\underline{P}(R)$. Since they are projective resolutions of M'' and M, respectively, there exists a chain map $\varphi: P \to P''$ covering $\eta: M \to M''$, where $P'' = \{P_i'', \partial_i''\}_{i \geq 0}$ and $P = \{P_j, \partial_j\}_{j \geq 0}$. The proof of (ii) in Theorem 5.28 implies that there is a resolution $0 \to M(\varphi)_\ell \to \cdots \to M(\varphi)_2 \to Z_1(M(\varphi)) \to M' \to 0$ where $\ell = \max(n, k+1)$. Moreover

$$0 \to Z_1(M(\varphi)) \to M(\varphi)_1 \to M(\varphi)_0 \to 0 \quad \text{is exact.} \tag{a}$$

Now

$$M(\varphi)_0 = P_0'' \in \underline{P}(R),$$

$$M(\varphi)_1 = P_1'' \oplus P_0 \in \underline{P}(R).$$

Thus sequence (a) splits. Hence $Z_1(M(\varphi))$ is projective and finitely generated, i.e., $Z_1(M(\varphi)) \in \underline{P}(R)$.

Clearly $M(\varphi)_i = P_i'' \oplus P_{i-1} \in \underline{P}(R)$ for all $i \geq 2$. This proves that M' is an FP-module.

(iii) Assume that M' and M are FP. Let

$$
\begin{array}{ccccccccc}
0 & \longrightarrow & P_m' & \longrightarrow & \cdots & \xrightarrow{\partial_1'} & P_0' & \xrightarrow{\varepsilon'} & M' & \longrightarrow & 0, \\
0 & \longrightarrow & P_k & \longrightarrow & \cdots & \xrightarrow{\partial_1} & P_0 & \xrightarrow{\varepsilon} & M & \longrightarrow & 0
\end{array}
$$

be resolutions of finite length with P_i', P_j in $\underline{P}(R)$. Let $P' = \{P_i', \partial_i'\}_{i \geq 0}$ and $P = \{P_i, \partial_i\}_{i \geq 0}$. There exists a chain map $\theta\colon P' \to P$ covering $j\colon M' \to M$. By (i) of Theorem 5.28, $M(\theta)$ is a resolution of M''. Now

$$M(\theta)_0 = P_0,$$

$$M(\theta)_i = P_i \oplus P_{i-1}' \quad \text{for } i \geq 1.$$

Thus each $M(\theta)_i$ is in $\underline{P}(R)$. Also the length of the complex $M(\theta) \leq \max(k, m + 1)$.

This proves that M'' is an FP-module. $\quad\square$

Corollary 5.29

If $0 \to M_n \xrightarrow{d_n} M_{n-1} \to \cdots \xrightarrow{d_1} M_0 \to 0$ is an exact sequence with all but one of the M_i's FP, then the remaining one also is FP.

Proof. Assume M_μ an FP module for $\mu \neq j$. Let $B_i = \operatorname{Im} d_{i+1}$ for $0 \leq i \leq n - 1$. Then $B_0 = M_0$ and we have exact sequences

$$\left.\begin{array}{c} 0 \to M_n \to M_{n-1} \to B_{n-2} \to 0 \\ \cdots \\ 0 \to B_{j+1} \to M_{j+1} \to B_j \to 0 \end{array}\right\}, \tag{a}$$

$$\left.\begin{array}{c} 0 \to B_{j-1} \to M_{j-1} \to B_{j-2} \to 0 \\ 0 \to B_{j-2} \to M_{j-2} \to B_{j-3} \to 0 \\ \cdots \\ 0 \to B_1 \to M_1 \to M_0 \to 0 \end{array}\right\}. \tag{b}$$

From the exact sequences (b) we see successively that $B_1, B_2, \ldots, B_{j-1}$ are all FP. Similarly from the exact sequences (a) we see successively that $B_{n-2}, \ldots, B_j$ are all FP. Now $0 \to B_j \to M_j \to B_{j-1} \to 0$ is exact. Hence M_j is an FP module. $\quad\square$

Corollary 5.30

If $\underline{\underline{H}}(R) = \underline{\underline{M}}(R)$, then R is left regular.

Proof. We have only to show that $\underline{\underline{H}}(R) = \underline{\underline{M}}(R) \Rightarrow R$ left noetherian. Let I be any left ideal of R. Since R and R/I are in $\underline{\underline{M}}(R)$ we get from

$\underline{\underline{H}}(R) = \underline{\underline{M}}(R)$ that R and R/I are FP. But $0 \to I \to R \to R/I \to 0$ is exact. Hence I is of type FP by Theorem 5.19. This proves that I is finitely generated. $\square$

Combining Proposition 5.10 and Theorem 5.28 we get the following result.

Corollary 5.31

Let $0 \to M' \xrightarrow{j} M \xrightarrow{\varphi} M'' \to 0$ be exact.

(i) If any two of the modules M', M, M'' have finite homological dimension, then the third one also has.

(ii) Let $d' = \operatorname{hd} M'$, $d = \operatorname{hd} M$, and $d'' = \operatorname{hd} M''$. Then (a) $d \leq \max(d', d'')$ (follows from Proposition 5.10), (b) $d' \leq \max(d'' - d, d)$ (follows from (ii) of Theorem 5.28) and (c) $d'' \leq \max(d, d' + 1)$ (follows from (i) of Theorem 5.28).

Chapter Four

Relationship between $\tilde{K}_0(R)$ and the Ideal Class Group When R Is a Dedekind Domain

Two good references for the material in this chapter are [5] and [29].

1 INTEGRAL ELEMENTS AND DEDEKIND DOMAINS

All the rings we consider in this section will be commutative with an identity element $1 \neq 0$. When we talk of a subring A of a ring B, it will be tacitly assumed that $1_A = 1_B$.

Let A be a subring of B and $x \in B$.

Definition 1.1

x is said to be integral over A if $\exists$ a monic polynomial $f(X) \in A[X]$ with $f(x) = 0$, namely if $\exists$ an integer $n \geq 1$ and elements $a_{n-1}, \ldots, a_0$ in A with $x^n + a_{n-1}x^{n-1} + \cdots + a_0 = 0$ in B.

Observe that every element $a \in A$ is integral over A since it is a root of the monic polynomial $X - a \in A[X]$.

Lemma 1.2

Let A be a subring of B, I an ideal of A, and M a f.g. A-module with $M \subset B$. Suppose $x \in B$ satisfies

(i) $xM \subset IM$,

(ii) $\{\lambda \in A[x] | \lambda M = 0\} = 0$.

Then $\exists$ a monic polynomial $f(X) = X^n + c_{n-1}X^{n-1} + \cdots + c_0 \in A[X]$ with all the $c_i \in I$ and $f(x) = 0$.

Proof. Let $u_1, \ldots, u_n$ generate M over A. Then $M = \sum_{i=1}^{n} Au_i$. From $xM \subset IM$ we get $xM \subset \sum_{i=1}^{n} Iu_i$. Hence we can find elements $a_{ji} \in I$ such that

$$xu_j = \sum_{i=1}^{n} a_{ji}u_i, \qquad 1 \le j \le n,$$

or

$$(x - a_{11})u_1 - a_{12}u_2 - \cdots - a_{1n}u_n = 0, \qquad (1)$$

$$- a_{21}u_1 + (x - a_{22})u_2 - \cdots - a_{2n}u_n = 0, \qquad (2)$$

$$\vdots$$

$$- a_{n1}u_1 - a_{n2}u_2 - \cdots + (x - a_{nn})u_n = 0. \qquad (n)$$

Let

$$\Delta = \det \begin{pmatrix} x - a_{11} & -a_{12} & \cdots & -a_{1n} \\ -a_{21} & x - a_{22} & \cdots & -a_{2n} \\ \vdots & \vdots & \ddots & \vdots \\ -a_{n1} & -a_{n2} & \cdots & x - a_{nn} \end{pmatrix}.$$

Let

$$\lambda_{ij} = -a_{ij} \quad \text{if } i \ne j,$$

$$\lambda_{ii} = x - a_{ii}, \qquad 1 \le i, \, j \ge n,$$

and Λ the matrix $(\lambda_{ij}) \in M_n(A[x])$. If α_{ij} is the cofactor of λ_{ij} in Λ, then we have

$$\sum_{i=1}^{n} \alpha_{ij}\lambda_{ij} = \Delta \quad \text{for any } j \text{ in } 1 \le j \le n$$

$$\sum_{i=1}^{n} \alpha_{ij}\lambda_{ik} = 0 \quad \text{if } 1 \le j \ne k \le n.$$

$\alpha_{11} \times (1) + \alpha_{21} \times (2) + \cdots + \alpha_{n1} \times (n)$ yields

$$\Delta u_1 = 0.$$

Similarly we get $\Delta u_2 = \cdots = \Delta u_n = 0$. Since $M = \sum_{i=1}^n A u_i$ it follows that $\Delta M = 0$. Now $\Delta \in A[x]$ and $\Delta \cdot M = 0$. From (ii) we see that $\Delta = 0$. The condition $\Delta = 0$ actually yields a relation of the form

$$x^n + c_{n-1}x^{n-1} + \cdots + c_1 x + c_0 = 0 \quad \text{with } c_i \in I.$$

This proves Lemma 1.2. □

Throughout Section 1, A will denote a subring of B.

Proposition 1.3

The following conditions on $x \in B$ are equivalent:

(i) x is integral over A.
(ii) *The subring $A[x]$ of B is f.g. as a module over A.*
(iii) ∃ *a subring R of B with $A[x] \subset R$ and R f.g. as a module over A.*
(iv) ∃ *a f.g. A-module $M \subset B$ with* (a) $xM \subset M$ *and* (b) $\{\lambda \in A[x] | \lambda \cdot M = 0\} = 0$.

Proof. (i) ⇒ (ii). If $x^n + a_{n-1}x^{n-1} + \cdots + a_0 = 0$ with $a_i \in A$, then $x^n = -\sum_{i=0}^{n-1} a_i x^i$ (with $x^0 = 1$). Thus $x^n \in \sum_{i=0}^{n-1} A x^i$. By induction on k we can now show that $x^{n+k} \in \sum_{i=0}^{n-1} A x^i$ for any $k \geq 0$. This proves that $1, x, \ldots, x^{n-1}$ generate $A[x]$ as a module over A.

(ii) ⇒ (iii). Trivial. Take $R = A[x]$.

(iii) ⇒ (iv). Also trivial. Take $M = R$. Since R is a ring and $x \in R$ we get $x \cdot R \subset R$ or $xM \subset M$. Moreover since $1 \in R = M$, $\lambda \cdot M = 0 \Rightarrow \lambda \cdot 1 = 0 \Rightarrow \lambda = 0$.

(iv) ⇒ (i). Is an immediate consequence of Lemma 1.2. □

Remark 1.4

If A is a subring of B, B a subring of C, and B is f.g. in A-mod, C is f.g. in B-mod, then C is f.g. in A-mod.

Proposition 1.5

Let $x_1, \ldots, x_k$ be finitely many elements in B, each integral over A. Then $A[x_1, \ldots, x_k]$ is f.g. in A-mod.

Proof. Induction on k. When $k = 1$, this is a consequence of (i) $\Rightarrow$ (ii) of Proposition 1.3. Let $k > 1$ and assume $A[x_1, \ldots, x_{k-1}]$ is f.g. in A-mod. Now x_k is integral over A and hence integral over $A[x_1, \ldots, x_{k-1}]$. Hence $A[x_1, \ldots, x_k] = A[x_1, \ldots, x_{k-1}][x_k]$ is f.g. in $A[x_1, \ldots, x_{k-1}]$-mod. By Remark 1.4, $A[x_1, \ldots, x_k]$ is f.g. in A-mod. $\square$

Corollary 1.6

The set $\overline{A} = \overline{A}_B = \{x \in B | x \text{ integral over } A\}$ is a subring of B satisfying $A \subset \overline{A}$.

Proof. Let x, y be any two elements of $\overline{A}$. Then from Proposition 1.5, $A[x, y]$ is f.g. in A-mod. We have $x - y \in A[x, y]$ and $x, y \in A[x, y]$. Hence $A[x - y] \subset A[x, y]$ and $A[xy] \subset A[x, y]$. From (i) $\Leftrightarrow$ (iii) of Proposition 1.3 we now see that $x - y$ and xy are integral over A. Clearly $\overline{A} \supset A$. This proves that $\overline{A}$ is a subring of B with $\overline{A} \supset A$. $\square$

Definition 1.7

(i) $\overline{A}$ is called the integral closure of A in B.

(ii) We say that A is integrally closed in B if $\overline{A} = A$.

(iii) If every element of B is integral over A, we say that B is integral over A.

Theorem 1.8

Let $A \subset B \subset C$ be a tower of rings. Suppose B is integral over A and C is integral over B. Then C is integral over A.

Proof. Let $x \in C$. Let $x^n + b_{n-1}x^{n-1} + \cdots + b_0 = 0$ with $b_i \in B$. Then x is integral over $A[b_0, \ldots, b_{n-1}]$. Hence, writing B' for $A[b_0, \ldots, b_{n-1}]$ we see that $B'[x]$ is f.g. in B'-mod. Also B' is f.g. in A-mod from Proposition 1.5. Hence $A[b_0, \ldots, b_{n-1}, x]$ is f.g. in A-mod. From (i) $\Rightarrow$ (iii) of Proposition 1.3 we now see that x is integral over A. $\square$

Let R be an integral domain and K its quotient field.

Definition 1.9

We say that R is integrally closed if R is integrally closed in K.

Definition 1.10

An integral domain R is called a *Dedekind domain* if it satisfies the following conditions:

(i) R is noetherian.

(ii) Every prime ideal $P \neq 0$ in R is a maximal ideal.

(iii) R is integrally closed.

Proposition 1.11

Any unique factorization domain (UFD) is integrally closed.

Proof. Let K be the quotient field of R. Let $\lambda = x/y \in K$, with $x \in R$ and $0 \neq y \in R$, be integral over R. We want to show that $\lambda \in R$. If $x = 0$, there is nothing to prove. If $x \neq 0$, we can assume that the only common divisors of x and y are units in R. Since λ is integral over R, we have a relation of the form

$$\left(\frac{x}{y}\right)^n + a_{n-1}\left(\frac{x}{y}\right)^{n-1} + \cdots + a_i\left(\frac{x}{y}\right) + a_0 = 0$$

with $a_i \in R$. Hence $x^n = -y(a_{n-1}x^{n-1} + a_{n-2}x^{n-2}y + \cdots + a_0 y^{n-1})$. Hence $y|x^n$ in R. If y is not a unit in R, this implies that $\exists$ a common irreducible factor of x and y, contradicting the assumption that the only common factors of x and y are units. Hence y is a unit in R. Then $\lambda = xy^{-1} \in R$. This proves Proposition 1.11. $\square$

We know that any principal ideal domain is a UFD. Moreover in a PID every prime ideal that is not equal to 0 is maximal. Clearly any PID is noetherian. Thus every PID is a Dedekind domain.

2 INTEGERS IN ALGEBRAIC NUMBER FIELDS

Definition 2.1

Any finite extension field K of the field Q of rational numbers is called an *algebraic number field*.

Definition 2.2

Any element $\alpha \in K$ that is integral over Z is called an *algebraic integer* in K.

From Corollary 1.6 we see that the algebraic integers in K form a subring, say R of K. Since K is a field, R is an integral domain. R is referred to as the ring of algebraic integers in K.

Remark 2.3

Since Z is a PID, from Proposition 1.11 we see that Z is integrally closed in Q. Thus the only rational numbers that are integral over Z are the elements of Z. In other words the subring of algebraic integers in Q is precisely Z. For this reason it is customary to refer to Z as the ring of rational integers.

Remark 2.4

Let A be a subring of B and $\overline{A}$ be the integral closure of A in B. Then $\overline{A}$ is integrally closed in B. In fact if $x \in B$ is integral over $\overline{A}$, since $\overline{A}$ is integral over A, it follows from Theorem 1.8 that x is integral over A and hence $x \in \overline{A}$.

Remark 2.5

Let K be an algebraic number field and R the ring of algebraic integers in K. Then K is the quotient field of R.

Let $0 \neq \gamma \in K$. Then γ is a root of some $f(X) \in Q[X]$. By multiplying by a suitable large integer we can assume $f(X) \in Z[X]$. Let $f(X) = a_n X^n + a_{n-1}X^{n-1} + \cdots + a_0$ with $a_i \in Z$ and $a_n \neq 0$. Since $a_n \gamma^n + a_{n-1}\gamma^{n-1} + \cdots + a_0 = 0$ we see that the element $a_n \gamma$ is a root of the monic polynomial

$$X^n + a_{n-1}X^{n-1} + a_{n-2}a_n X^{n-2} + \cdots + a_0 a_n^{n-1} = 0$$

whose coefficients are clearly in Z. Thus $a_n \gamma \in R$. Thus we have shown that any $0 \neq \gamma \in K$ can be written as $a_n^{-1} \cdot a_n \gamma = (a_n \gamma)|a_n$ with $a_n \gamma \in R$ and $a_n \in Z$, i.e., $\gamma = \alpha|k$ with $\alpha \in R$ and $0 \neq k \in Z$.

Remark 2.6

For any $\alpha \in K$ where K is an algebraic number field we write Irr(α, Q) the irreducible polynomial of α over Q, taken with leading coefficient 1.

Then α is an algebraic integer in $K \Leftrightarrow$ Irr$(\alpha, Q) \in Z[X]$.

Proof. If Irr$(\alpha, Q) \in Z[X]$, then α is a root of the monic polynomial Irr(α, Q) with coefficients in Z. Hence α is an algebraic integer.

Conversely, suppose α is an algebraic integer. Then $\exists$ a monic polynomial $h(X) \in Z[X]$ with $h(\alpha) = 0$. Then $h(X) = f(X)g(X)$ in $Q[X]$ where $f(X) =$ Irr(α, Q). Now $f(X) \in Q[X]$ and $g(X) \in Q[X]$. We can write $f(X) = [cF(X)]/d$, $g(X) = [eG(X)]/\ell$ with c, d, e, ℓ integers greater than or equal to 1, $F(X), G(X)$ primitive in $Z[X]$, i.e., to say g.c.d. of coefficients of

$F(X) = 1 =$ g.c.d. of coefficients of $G(X)$. Then $ceF(X)G(X) = d\ell h(X)$. Since the product of primitive polynomials is primitive, $F(X)G(X)$ is primitive. Also $h(X) \in Z[X]$ being monic, $h(X)$ is primitive. It follows that $ce = d\ell$ and $F(X)G(X) = h(X)$. In particular, $F(X) \in Z[X]$ is monic. From $f(X) = [cF(X)]/d$ with leading coefficient of $f(X) = 1$, leading coefficient of $F(X) = 1$, we see that $c = d$. Hence $f(X) = F(X) \in Z[X]$. Thus α algebraic integer $\Rightarrow \mathrm{Irr}(\alpha, Q) \in Z[X]$. $\square$

Theorem 2.7

Let K be an algebraic number field with $(K : Q) = n$. Let R be the ring of algebraic integers in K. Then R as a Z-module is free of rank n.

Proof. Since char $Q = 0$, we see that K is a separable extension of Q. Hence $\exists$ an element $\gamma \in K$ with $K = Q(\gamma)$. From Remark 2.5 we see that $\gamma = \alpha/a$ with $\alpha \in R$ and $0 \neq a \in Z$. Then $K = Q(\gamma) = Q(\alpha)$ (since $a \in Z \subset Q$). Since $(Q(\alpha) : Q) = n$ it follows that $1, \alpha, \alpha^2, \ldots, \alpha^{n-1}$ form a basis of $Q(\alpha)$ over Q as a vector space.

Let β be any element of R. Since $R \subset K$ we can write $\beta = \sum_{i=0}^{n-1} \lambda_i \alpha^i$ with $\lambda_i \in Q$ and uniquely determined by β. The minimum polynomial of α, necessarily of degree n, has all its roots distinct. Let $\alpha = \alpha_1, \alpha_2, \ldots, \alpha_n$ be the distinct conjugates of α lying in a finite normal extension L of Q. For instance L could be chosen to be a splitting field of $\mathrm{Irr}(\alpha, Q)$ containing the element α. Then all the conjugates of β over Q are given by

$$\beta_j = \sum_{i=0}^{n-1} \lambda_i \alpha_j^i, \qquad 1 \leq j \leq n,$$

but the elements $\beta_1, \ldots, \beta_n$ need not all be distinct. [For instance if $\omega = e(2\pi\sqrt{-1}/3)$ is a primitive cube root of unity, ω, ω^2 are conjugates of ω over Q, but for $\beta = \omega + \omega^2$ there is only one conjugate over Q and it is β itself. In fact $\beta = -1$.]

Let

$$A = \begin{pmatrix} 1 & \alpha_1 & \alpha_1^2 & \cdots & \alpha_1^{n-1} \\ 1 & \alpha_2 & \alpha_2^2 & \cdots & \alpha_2^{n-1} \\ \vdots & \vdots & \vdots & \ddots & \vdots \\ 1 & \alpha_n & \alpha_n^2 & \cdots & \alpha_n^{n-1} \end{pmatrix}.$$

Then $\det A = \prod_{1 \leq i < j \leq n} (\alpha_i - \alpha_j) \neq 0$ in L. Now L/Q is a Galois extension and any $\sigma \in \mathrm{Gal}(L/Q)$ permutes the α_j's. Hence $c = (\det A)^2$ is left fixed by every element $\sigma \in \mathrm{Gal}(L/Q)$. It follows that $c = (\det A)^2 \in Q$.

Since α is an integer in K, it follows that all the elements α_j^i are algebraic integers. det A is a polynomial in the various α_j^i with coefficients from Z. Hence det A and therefore $c = (\det A)^2$ are both algebraic integers. Since Z is integrally closed it follows that $c \in Z$.

Using Cramér's method we solve the system

$$\sum_{i=0}^{n-1} \lambda_i \alpha_j^i = \beta_j, \qquad 1 \le j \le n, \qquad (*)$$

of linear equations for the λ_i's over the field L. We get

$$\lambda_0 = \det \frac{\begin{pmatrix} \beta_1 & \alpha_1 & \alpha_1^{n-1} \\ \vdots & \vdots & \vdots \\ \beta_n & \alpha_n & \alpha_n^{n-1} \end{pmatrix}}{\det A}, \dots, \lambda_{n-1} = \det \frac{\begin{pmatrix} 1 & \alpha_1 & \alpha_1^{n-2} & \beta_1 \\ 1 & & & \\ 1 & \alpha_n & \alpha_n^{n-2} & \beta_n \end{pmatrix}}{\det A}$$

or

$$\lambda_i = |A|^{-1} \sum_{j=1}^{n} \gamma_{ij} \beta_j$$

where $|A| = \det A$ and γ_{ij} are polynomials in α_μ^ν's with coefficients from Z. In particular γ_{ij} are algebraic integers.

We started with β in R. Thus $\beta_1 = \beta$ is an algebraic integer. Hence $\mathrm{Irr}(\beta, Q) \in Z[X]$. Thus every conjugate of β being a root of $\mathrm{Irr}(\beta, Q)$ is also an algebraic integer. Hence β_j is an algebraic integer for all $1 \le j \le n$. Now $c\lambda_i = |A| \sum_{j=1}^n \gamma_{ij} \beta_j$. Since $|A|$, β_j, and γ_{ij} are all algebraic integers we see that each $c\lambda_i$ ($0 \le i \le n - 1$) is an algebraic integer.

But $c \in Z$ and $\lambda_i \in Q$. Thus $c\lambda_i \in Q$ is algebraic integer or $c\lambda_i \in Z$. Let $c\lambda_i = m_i \in Z$. Then $\beta = (1/c)\sum_{i=0}^{n-1} m_i \alpha^i$ or $\beta \in (1/c)Z[\alpha]$. Here c is independent of β and β is arbitrary in R. Thus $R \subset (1/c)Z[\alpha]$. Since $\alpha \in R$, we see that $Z[\alpha]$ is f.g. in Z-mod. But $(1/c)Z[\alpha] \simeq Z[\alpha]$ (in Z-mod), hence f.g. in Z-mod.

Since Z is a noetherian ring, it follows that R is finitely generated over Z. Since R is an integral domain containing Z, it is clear that R is torsion free in Z-mod.

Hence R is free of finite rank over Z. Let $x_1, \dots, x_k$ be a basis for R over Z. Then it is trivially seen that $x_1, \dots, x_k$ is a basis for K over Q. Since $(K:Q) = n$, we get $k = n$. $\square$

Theorem 2.8

Let K be an algebraic number field and R the ring of algebraic integers in K. Then any nonzero prime ideal P in R is maximal.

Proof. Let $0 \neq \alpha \in P$. Since α is an integer in K we see that $\mathrm{Irr}(\alpha, Q) \in Z[X]$. Let $X^m + a_{m-1}X^{m-1} + \cdots + a_0$ be the irreducible polynomial of α. Then $a_0 \neq 0$ and as commented already $a_i \in Z$ for all i. We have $\alpha^m + a_{m-1}\alpha^{m-1} + \cdots + a_0 = 0$. Hence $a_0 = -\alpha(\alpha^{m-1} + a_{m-1}\alpha^{m-2} + \cdots + a_1) \in \alpha R \subset P$. Thus $0 \neq a_0 \in P \cap Z$. It is clear that $P \cap Z$ is an ideal in Z. Moreover it is a prime ideal in Z. In fact if m, n are elements of Z with $mn \in P \cap Z$ and $m \notin P \cap Z$, then $m \notin P$. Hence $n \in P$ or $n \in P \cap Z$. It follows that $P \cap Z = pZ$ for a prime $p \in Z$. In particular $P \cap Z$ is a maximal ideal in Z.

Let A be an ideal of R with $P \subseteq A \subsetneq R$. Then $A \cap Z$ is an ideal of Z with $A \cap Z \subset P \cap Z = pZ$. Moreover $1 \notin A$. Hence $A \cap Z \neq Z$. It follows that $A \cap Z = pZ$. We want to show that $A = P$. If this is not the case, there will exist a $\beta \in A$ with $\beta \notin P$. Then from $\beta \in R$ we see that $\mathrm{Irr}(\beta, Q) \in Z[X]$. Let $X^n + b_{n-1}\beta^{n-1} + \cdots + b_0 = 0$ and we get $b_0 = -\beta(\beta^{n-1} + b_{n-1}\beta^{n-2} + \cdots + b_1) \in A$. Thus $b_0 \in A \cap Z = P \cap Z$. Hence $\beta(\beta^{n-1} + b_{n-1}\beta^{n-2} + \cdots + b_1) \in P$. Since $\beta \notin P$ and P is a prime ideal in R, we see that $\beta^{n-1} + b_{n-1}\beta^{n-2} + \cdots + b_1 \in P$. Now $\beta^{n-1} + b_{n-1}\beta^{n-2} + \cdots + b_1 = \beta(\beta^{n-2} + b_{n-1}\beta^{n-3} + \cdots + b_2) + b_1 \in P \subset A$. Since $\beta(\beta^{n-2} + b_{n-1}\beta^{n-3} + \cdots + b_2)$ is in $\beta R \subset A$, we see that $b_1 \in A$. Hence $b_1 \in A \cap Z = P \cap Z$. In particular $b_1 \in P$. It follows that $\beta(\beta^{n-2} + b_{n-1}\beta^{n-3} + \cdots + b_2) \in P$. Since $\beta \notin P$, we get $\beta^{n-2} + b_{n-1}\beta^{n-3} + \cdots + b_2 \in P$. Iterating this argument, we obtain successively that $\beta^{n-3} + b_{n-1}\beta^{n-4} + \cdots + b_3 \in P, \ldots, \beta^2 + b_{n-1}\beta + b_{n-2} \in P$, $\beta + b_{n-1} \in P$, and $\beta \in P$. This contradicts the assumption $\beta \notin P$. Hence $P = A$ and P is a maximal ideal in R. $\square$

Remark 2.9

Let R denote the ring of algebraic integers in an algebraic number field K. Then from Remark 2.5 we see that K is the quotient field of R. From Theorem 1.8 we see that any element α of K which is integral over R will be integral over Z and hence $\alpha \in R$. Thus R is integrally closed in its quotient field K. From Theorem 2.8 we see that R is f.g. even as a Z-module and hence is a quotient of $Z[X_1, \ldots, X_n]$ as a ring for some integer $n \geq 1$. Hence R is noetherian. Theorem 2.8 asserts that any nonzero prime ideal of R is maximal. Thus R is a Dedekind domain.

3   FRACTIONAL IDEALS AND THE CLASS GROUP

Let R be an integral domain and K its quotient field. Let R^* (resp. K^*) denote the set of nonzero elements of R (resp. K).

Definition 3.1

A *fractional ideal* of R is a *nonzero* R-submodule A of K with the property that $\exists$ some $d \in R^*$ with $dA \subseteq R$.

In this case dA is a nonzero R-submodule of R and hence either $dA = R$ or dA is a nonzero ideal in R.

Definition 3.2

A fractional ideal A or R will be called *integral* if $A \subseteq R$. Clearly R and all nonzero ideals in R are exactly the integral ideals of R.

Since K is a field, for any $\lambda \in K^*$ we have $K\lambda = K$. Hence for any fractional ideal A of R we have $KA = K$. For any $\lambda \in K^*$, $R\lambda$ is clearly a fractional ideal of R. In fact if $\lambda = x/y$ with $x \in R^*$ and $y \in R^*$, then $yR\lambda \subseteq R$.

Definition 3.3

Any fractional ideal of R of the form $R\lambda$, with $\lambda \in K^*$ is called a *principal fractional ideal*.

Let J denote the set of fractional ideals of R. For any A, B in J it is easy to see that $A \cap B$, $A + B$, and AB are all in J (observe that $AB = \{$finite sums $\sum a_i b_i | a_i \in A,\ b_i \in B \})$. In fact if d, d' in R^* satisfy $dA \subseteq R$, $d'B \subseteq R$, we have

$$dd'(A + B) \subseteq R, \qquad dd'(AB) \subseteq R,$$

$$d(A \cap B) \subseteq dA \subseteq R,$$

$$d'(A \cap B) \subseteq d'B \subseteq R.$$

Clearly, $A + B \supseteq A \neq 0$ and $AB \neq 0$. Also if $0 \neq \lambda = x/y \in A$ and $0 \neq \mu = u/v \in B$ with x, y, u, v in R^* we have $0 \neq xu \in A \cap B$.

Definition 3.4

For any A, B in J we define $(A : B) = \{\lambda \in K | \lambda B \subseteq A\}$.

Clearly $(A : B)$ is an R-submodule of K. Let d, d' in R^* satisfy $dA \subset R$ and $d'B \subset R$.

If $0 \neq a \in A$, from $d'B \subset R$ we get $ad'B \subset A$ and hence $0 \neq ad' \in (A : B)$. Also if $0 \neq b \in B$, then $0 \neq r = d'b \in R \cap B$. Moreover

$$dr(A : B) = d(A : B)r \subseteq dA \quad \text{since } r \in B \subseteq R.$$

Thus $dr \in R^*$ satisfies $dr(A : B) \subseteq R$. Hence $(A : B) \in J$.

It is clear that J is a commutative monoid under multiplication $A \cdot B = AB$ with R as the identity element. We will address ourselves to the question: When is J a group?

Definition 3.5

A fractional ideal A of R is said to be *invertible* if it is invertible as an element of the monoid J. Namely A is invertible if $\exists$ an $A' \in J$ with $AA' = R$.

Clearly, every principal fractional ideal $R\lambda$ with $\lambda \in K^*$ is invertible with $R\lambda^{-1}$ as its inverse.

We know that if an element in a monoid is invertible, then its inverse is unique. In case $A \in J$ is invertible we write A^{-1} for its inverse in J.

Lemma 3.6

If $A \in J$ is invertible, then $A^{-1} = (R : A)$.

Proof. From $A^{-1}A = R$ we see that

$$A^{-1} \subseteq (R : A). \tag{i}$$

Also $(R : A)A \subseteq R$. Hence

$$(R : A)AA^{-1} \subseteq RA^{-1} = A^{-1} \quad \text{or} \quad (R : A)R \subseteq A^{-1} \quad \text{or} \quad (R : A) \subseteq A^{-1}. \tag{ii}$$

From (i) and (ii) we get $(R : A) = A^{-1}$. $\square$

Caution. A^{-1} should not be confused with $\{\lambda^{-1} | \lambda \in A\}$. In fact for any $A \in J$ we have $0 \in A$ and 0^{-1} does not make sense.

If $A = R\lambda$ with $\lambda \in K^*$, then A is invertible in J and $A^{-1} = R\lambda^{-1}$.

Lemma 3.7

Suppose $A \in J$ is invertible. Then A is f.g. in R-mod.

Proof. From $A \cdot A^{-1} = R$ we get $1 = \sum_{i=1}^{k} x_i x_i'$ with $x_i \in A$ and $x_i' \in A^{-1}$. Then for any $a \in A$ we have $x_i' a \in R$ and hence $a = \sum_{i=1}^{k} x_i x_i' a \in \sum_{i=1}^{k} Rx_i$. Thus $A \subseteq \sum_{i=1}^{k} Rx_i$. But $x_i \in A \Rightarrow \sum_{i=1}^{k} Rx_i \subseteq A$. Thus $A = \sum_{i=1}^{k} Rx_i$. $\square$

Observe that J is a group $\Leftrightarrow$ every $A \in J$ is invertible.

Lemma 3.8

If every integral ideal of R is invertible, then every fractional ideal of R is invertible. Hence J is a group $\Leftrightarrow$ every integral ideal of R is invertible.

Proof. Let A be a fractional ideal of R. Then $\exists$ a $d \in R^*$ with dA an integral ideal. If $C = dA$ and C^{-1} is the inverse of C (which exists by assumption), then

$$A \cdot dC^{-1} = dA \cdot C^{-1} = C \cdot C^{-1} = R.$$

Hence dC^{-1} is the inverse of A. $\square$

Lemma 3.9

Let R be a noetherian integral domain with K as its quotient field. Then $A \subset K$ is a fractional ideal of $R \Leftrightarrow A$ is a nonzero finitely generated R-submodule of K.

Proof. Let A be a fractional ideal of R. Then $A \neq 0$ and $\exists$ a $d \in R^*$ with $dA \subset R$. Since R is noetherian, dA is f.g. in R-mod. But $A \simeq dA$ in R-mod. Hence A is f.g. in R-mod.

Conversely, let $A \neq 0$ be a f.g. R-submodule of K. Let $x_2, \ldots, x_k$ generate A over R. If $x_k = a_k/d_k$ with $a_k \in R$ and $d_k \in R^*$, then $dA \subset R$ where $d = d_1 d_2 \cdots d_k$. $\square$

For the *remainder of this section R will denote a Dedekind domain with quotient field K*. J will denote the set of fractional ideals of R. Since R is noetherian, from Lemma 3.9 we see that $A \in J \Leftrightarrow A$ is a nonzero f.g. R-submodule of K. For any $A \in J$, one has $KA = K$. For the sake of having a meaningful decomposition theory for nonzero ideals in R, we will allow R to be considered as a prime ideal of R.

Lemma 3.10

Any integral ideal A of R contains a product of nonzero prime ideals in R.

Proof. Let $\mathscr{F} = \{ I | I$ an integral ideal of R, I does not contain any product of nonzero prime ideals of $R \}$. To prove Lemma 3.10 we have only to prove that $\mathscr{F}$ is empty. Suppose $\mathscr{F} \neq \varnothing$. Then since R is noetherian, $\exists$ a maximal element M in $\mathscr{F}$. If $M = R$, then R is a prime ideal in M. Hence $M \neq R$. Also M itself is not prime. Let α, β be elements of R with $\alpha \notin M$, $\beta \notin M$, and $\alpha\beta \in M$. Then $M + R\alpha \supsetneq M$, $M + R\beta \supsetneq M$. Clearly $M + R\alpha$ and $M + R\beta$ are not in $\mathscr{F}$ because of the maximality of M in $\mathscr{F}$. Thus $M + R\alpha \supset P_1 \cdots P_k$ and $M + R\beta \supset P_1' \cdots P_\ell'$ where P_i, P_j' are nonzero prime ideals of R. Hence $(M + R\alpha) \cdot (M + R\beta) \supset P_1 \cdots P_k P_1' \cdots P_\ell'$. Moreover $(M + R\alpha) \cdot (M + R\beta) \subset M + R\alpha \cdot R\beta \subseteq M$. Thus $M \supset P_1 \cdots P_k P_1' \cdots P_\ell'$, a contradiction. Hence $\mathscr{F} = \varnothing$. □

Lemma 3.11

Let A be a proper ideal of R (i.e., $0 \neq A \subsetneq R$). Then

$$(R : A) = \{\lambda \in K | \lambda A \subset R\} \supsetneq R.$$

Proof. From $RA \subseteq R$ we immediately get

$$(R : A) \supseteq R. \tag{i}$$

Let $0 \neq \alpha \in A$. From Lemma 3.10 we see that $A \supset R\alpha \supset P_1 \cdots P_n$ with P_i nonzero prime ideals of R. Pick such a set of prime ideals with minimal possible n. ∃ a maximal ideal P of R with $A \subseteq P$. Hence $P \supset P_1 \cdots P_n$. If $P_i \subsetneq P$ for every i, then ∃ an $\alpha_i \in P_i$ with $\alpha_i \notin P$. Then $\alpha_1 \cdots \alpha_n \in P$ with $\alpha_i \notin P$ for every i contradicting the prime nature of P. Hence $P_i \subseteq P$ for some i, say $P \supseteq P_1$. Since every prime ideal in R is maximal, we see that $P = P_1$. Thus

$$PP_2 \cdots P_n \subset R\alpha \subset A \subset P.$$

By the minimality of n we have $P_2 \cdots P_n \subsetneq R\alpha$. Let $\beta \in P_2 \cdots P_n$ with $\beta \notin R\alpha$. Let $\lambda = \alpha^{-1}\beta \in K$. Then $\lambda \notin R$. Moreover

$$\lambda A = \alpha^{-1}\beta A \subseteq \alpha^{-1}P_2 \cdots P_n A \subseteq \alpha^{-1}P_2 \cdots P_n P$$

$$\subseteq \alpha^{-1}PP_2 \cdots P_n$$

$$\subseteq R \quad \text{since } PP_2 \cdots P_n \subseteq R\alpha.$$

Thus $\lambda \in (R : A)$ and $\lambda \notin R$. This completes the proof of Lemma 3.11. □

Lemma 3.12

If A is an integral ideal of R, then $(R : A)A = R$.

Proof. Let $B = (R : A)A$. From the definition of $(R : A)$ we have $(R : A)A \subseteq R$. Thus $B \in J$ and $B \subseteq R$. Hence B is an integral ideal of R. Hence $(R : B) \in J$ and $B(R : B) \subseteq R$. Thus $A(R : A)(R : B) = B(R : B) \subseteq R$. Thus $(R : A)(R : B) \subseteq (R : A)$. For any $\beta \in (R : B)$ we have $(R : A)\beta \subset (R : A)$. By induction on n we get $(R : A)\beta^n \subset (R : A)$ for all $n \geq 1$. Thus $(R : A)[\beta]$

is an R-submodule of $(R:A)$ and $(R:A) \in J$ and hence f.g. over R. Since R is noetherian, $(R:A)[\beta]$ is f.g. over R. Now $R[\beta]$ is an R-submodule of $(R:A)[\beta]$ and hence $R[\beta]$ is f.g. in R-mod. Thus β is integral over R. Since R is integrally closed we get $\beta \in R$. Thus $(R:B) = R$. Lemma 3.11 now implies $B = R$. Thus

$$(R:A)A = R. \qquad \square$$

Corollary 3.13

When R is a Dedekind domain, Lemma 3.12 shows that every integral ideal of R is invertible. Also, Lemma 3.8 shows that every fractional ideal will then be invertible and hence J is a group. From Lemma 3.6 for any $C \in J$ we have $C^{-1} = (R:C)$.

Theorem 3.14

Every proper ideal A in R is a product of proper prime ideals. Moreover if $A = P_1 \cdots P_k = P_1' \cdots P_\ell'$ with P_i, P_j' proper prime ideals, then $k = \ell$ and there exists a permutation σ of $\{1, 2, \ldots, k\}$ with $P_i = P_{\sigma(i)}'$.

Proof. By Lemma 3.10, $A \supset L_1 \cdots L_c$ with each L_i a nonzero proper prime ideal of R. By induction on c we will prove that any ideal that contains a product of c proper prime ideals is itself a product of proper prime ideals. Let $c = 1$. Then $A \supset L_1$. Since every proper prime ideal is maximal in R, we get $A = L_1$. Thus the statement is valid for $c = 1$.

Let $c > 1$. $\exists$ a maximal ideal $P \supset A$ in R. Then $P \supset A \supset L_1 \cdots L_c$. The argument used in the proof of Lemma 3.11 shows that $P = L_i$ for some i, say $P = L_1$. Then $R = P^{-1}P \supset P^{-1}A \supset P^{-1}L_1L_2 \cdots L_c = L_2 \cdots L_c$. Thus $P^{-1}A$ is a proper ideal of R (if $P^{-1}A = R$, then $A = P$ is a proper prime ideal, hence we can assume $P^{-1}A \neq R$) containing a product of $(c-1)$ proper prime ideals. Hence $P^{-1}A = Q_1 \cdots Q_r$ with each Q_r a proper ideal in R. Hence

$$A = PQ_1 \cdots Q_r. \qquad \square$$

The uniqueness part of Theorem 3.14 follows from Lemma 3.15.

Lemma 3.15

Let $P_1 \cdots P_n \subset P_1' \cdots P_m'$ with P_i, P_j' proper prime ideals in R. Then $m \leq n$ and there is an injection $\theta \colon \{1, \ldots, m\} \to \{1, \ldots, n\}$ such that

$$P_j' = P_{\theta(j)}.$$

Proof. We have $P_1 \cdots P_n \subset P_1' \cdots P_m' \subset P_1'$. The argument used in the proof of Lemma 3.11 shows that $P_1' = P_i$ for some i, say $P_1' = P_1$. Then multiplying by P_1^{-1} we get

$$P_2 \cdots P_n \subset P_2' \cdots P_m'.$$

Iteration of the preceding argument now completes the proof. □

From now on, by a prime ideal we mean a proper prime ideal in R. Then any integral ideal $A \neq R$ can be uniquely written as $P_1^{r_1} \cdots P_n^{r_n}$ with each r_i an integer greater than 0 and $P_1, \ldots, P_n$ distinct prime ideals. We will agree to write $R = P^0$ for any prime ideal P in R.

Corollary 3.16

Let A, B be integral ideals of R. Then $A \subset B \Leftrightarrow A = BC$ for some integral ideal C of R.

Proof. Suppose $A = BC$ with C an integral ideal. Then $A = BC \subset BR \subset B$. Conversely, assume $A \subset B$. If $B = R$ choose $C = A$. Then $A = RC$. Suppose $B \neq R$. Then $A = P_1 \cdots P_n$ and $B = P_1' \cdots P_m'$ with P_i, P_j' prime ideals in R. From Lemma 3.15 we have $m \leq n$ and rearranging the P_i's if necessary we may assume $P_j = P_j'$ for $1 \leq j \leq m$. Thus $C = P_{m+1} \cdots P_n$ satisfies $A = BC$. □

Let $A \in J$. Then $\exists$ a $d \in R^*$ with $dA \subset R$. Let $dA = P_1^{r_1} \cdots P_n^{r_n}$ with $r_i > 0$ and $P_1, \ldots, P_n$ distinct primes in R. Also let $Rd = Q_1^{s_1} \cdots Q_k^{s_k}$ with $Q_1, \ldots, Q_k$ distinct primes and $s_j > 0$. Then $A = (Rd)^{-1} dA = P_1^{r_1} \cdots P_n^{r_n} Q_1^{-s_k} \cdots Q_k^{-s_k}$. If any P_i is a certain Q_j, then $P_i^{r_i} Q_j^{-s_j} = P_i^{r_i - s_j}$. Thus we can write A as $C_1^{\lambda_1} \cdots C_\ell^{\lambda_\ell}$ with $C_1, \ldots, C_\ell$ distinct prime ideals in R and $\lambda_i \in Z^*$ (i.e., $\lambda_i \neq 0$ an integer) and this expression is unique. In other words J is a free abelian group with the set of prime ideals in R as a basis.

Theorem 3.17

Let A, B be integral ideals. Suppose

$$A = P_1^{a_1} \cdots P_n^{a_n},$$

$$B = P_1^{b_1} \cdots P_n^{b_n},$$

with a_i, b_i integers greater than or equal to 0 and $P_1, \ldots, P_n$ distinct prime ideals in R. Then

(i) $A + B = P_1^{c_1} \cdots P_n^{c_n}$ *where $c_i = \min(a_i, b_i)$,*
(ii) $A \cap B = P_1^{d_1} \cdots P_n^{d_n}$ *where $d_i = \max(a_i, b_i)$.*

Proof. Since $c_i \leq a_i$ we have $A \subset P_1^{c_1} \cdots P_n^{c_n}$. Similarly $c_i \leq b_i \Rightarrow B \subset P_1^{c_1} \cdots P_n^{c_n}$. Hence

$$P_1^{c_1} \cdots P_n^{c_n} \supset A + B. \qquad (a)$$

Let $A + B = P_1^{\lambda_1} \cdots P_n^{\lambda_n}$ be the factorization of $A + B$. In fact we could assume such a factorization for $A + B$, because we could introduce 0th powers of prime ideals in any product and hence the factorizations of A and B could be assumed to contain all prime ideals occurring in the factorization of $A + B$ at least to the 0th power.

From $A \subset P_1^{\lambda_1} \cdots P_n^{\lambda_n}$ and Lemma 3.15 we see that $\lambda_i \leq a_i$. Similarly $B \subset P_1^{\lambda_1} \cdots P_n^{\lambda_n} \Rightarrow \lambda_i \leq b_i$. Thus $\lambda_i \leq \min(a_i, b_i) = c_i$. Hence $A + B = P_1^{\lambda_1} \cdots P_n^{\lambda_n} \supset P_1^{c_1} \cdots P_n^{c_n}$. Thus $A + B = P_1^{\lambda_1} \cdots P_n^{\lambda_n} \supset P_1^{c_1} \cdots P_n^{c_n} \supset A + B$ or $P_1^{c_1} \cdots P_n^{c_n} = A + B$. This proves (i).

The proof of (ii) is similar. $\square$

Corollary 3.18

For any two integral ideals A, B of R we have $AB = (A + B) \cdot (A \cap B)$.

Definition 3.19

Two integral ideals A, B are said to be relatively prime if $A + B = R$.

Corollary 3.20

Let $A = P_1^{\lambda_1} \cdots P_n^{\lambda_n}$ with $\lambda_i > 0$ for each i, $P_1, \ldots, P_n$ distinct prime ideals; $B = Q_1^{\mu_1} \cdots Q_\ell^{\mu_\ell}$ with $\mu_j > 0$ for each j and $Q_1, \ldots, Q_\ell$ distinct prime ideals. Then A, B are relatively prime $\Leftrightarrow P_i \neq Q_j$ for any (i, j).

Corollary 3.21

If A, B are relatively prime ideals in R, then $A \cap B = A \cdot B$.

Proof. Immediate from Corollary 3.18. $\square$

Corollary 3.22

Integral ideals $A_1, \ldots, A_k$ are pairwise relatively prime $\Leftrightarrow A_i + \prod_{j \neq i} A_j = R$ $(1 \leq i \leq k)$.

Proof. Let $A_i = P_1^{\lambda_{i_1}} P_2^{\lambda_{i_2}} \cdots P_r^{\lambda_{i_r}}$ $(1 \leq i \leq k)$, with $P_1, \ldots, P_r$ district prime ideals in R and $\lambda_{i\mu} \geq 0$ for all $1 \leq \mu \leq r$, $1 \leq i \leq k$. Then $A_1, \ldots, A_k$ are pairwise relatively prime $\Leftrightarrow$ for any $1 \leq i \neq j \leq k$, $\min(\lambda_{i\mu}, \lambda_{j\mu}) = 0$ for

$1 \leq \mu \leq r$. Also

$$\prod_{j \neq i} A_j = P_1^{\Sigma_{j \neq i} \lambda_{j1}} P_2^{\Sigma_{j \neq i} \lambda_{j2}} \cdots P_r^{\Sigma_{j \neq i} \lambda_{jr}},$$

$$A_i = P_1^{\lambda_{i1}} \cdots P_r^{\lambda_{ir}}.$$

Hence $A_i + \prod_{j \neq i} A_j = R \Leftrightarrow \min(\lambda_{i\mu}, \Sigma_{j \neq i} \lambda_{j\mu}) = 0$ for $1 \leq \mu \leq r$.

Let μ be a given number with $1 \leq \mu \leq r$. Now, if $\lambda_{i\mu} \neq 0$, then $\min(\lambda_{i\mu}, \lambda_{j\mu}) = 0$ for all $j \neq i \Leftrightarrow$ each $\lambda_{j\mu} = 0$ for $j \neq i$; $\Leftrightarrow \min(\lambda_{i\mu}, \Sigma_{j \neq i} \lambda_{j\mu}) = 0$. If $\lambda_{i\mu} = 0$, then clearly $\min(\lambda_{i\mu}, \lambda_{j\mu}) = 0$ for all $j \neq i$. Hence $A_1, \ldots, A_k$ are pairwise relatively prime

$$\Leftrightarrow \quad A_i + \prod_{j \neq i} A_j = R, \quad 1 \leq i \leq k. \qquad \square$$

Corollary 3.23

Let $A_1, \ldots, A_n$ be pairwise relatively prime integral ideals in R. Then

$$A_1 \cap \cdots \cap A_n = A_1 A_2 \cdots A_n.$$

Proof. For $n = 2$ this is Corollary 3.21. Let $n > 2$ and by induction assume

$$A_1 \cap \cdots \cap A_{n-1} = A_1 A_2 \cdots A_{n-1}.$$

From Corollary 3.22 we have

$$A_n + A_1 A_2 \cdots A_{n-1} = R.$$

Hence Corollary 3.21 yields

$$A_n \cap (A_1 A_2 \cdots A_{n-1}) = A_1 A_2 \cdots A_{n-1} A_n$$

or

$$A_n \cap (A_1 \cap A_2 \cap \cdots \cap A_{n-1}) = A_1 A_2 \cdots A_n. \qquad \square$$

Theorem 3.24 (Chinese Remainder Theorem)

Let $A_1, \ldots, A_n$ be pairwise relatively prime integral ideals of R. Given $\alpha_i \in R$ $(1 \leq i \leq n)$ $\exists$ an $\alpha \in R$ with $\alpha \equiv \alpha_i \pmod{A_i}$, $1 \leq i \leq n$. Moreover α is uniquely determined modulo $A_1 A_2 \cdots A_n$.

Proof. Since $A_i + \prod_{j \neq i} A_j = R$, there exist $\beta_i \in A_i$, $\beta_i' \in \prod_{j \neq i} A_j$, with $1 = \beta_i + \beta_i'$. Let $\alpha = \sum_{i=1}^n \alpha_i \beta_i'$. Then

$$\alpha \equiv \alpha_i \beta_i' \pmod{A_i} \quad \text{since } \alpha_j \beta_j' \in A_i \text{ whenever } j \neq i$$

$$\equiv \alpha_i \pmod{A_i} \quad \text{since } \beta_i' = 1 - \beta_i \equiv 1 \pmod{A_i}.$$

Thus $\alpha \equiv \alpha_i \pmod{A_i}$.

Suppose α, α' are any two elements of R with $\alpha \equiv \alpha_i \pmod{A_i} \equiv \alpha'$. Then $\alpha - \alpha' \in A_i$ for all i or $\alpha - \alpha' \in A_1 \cap \cdots \cap A_n = A_1 A_2 \cdots A_n$. □

Corollary 3.25

Let $A_1, \ldots, A_n$ be pairwise relatively prime, proper integral ideals in R and $\eta_i : R \to R/A_i$ the canonical quotient map. Then

$$\varphi : \frac{R}{A_1 \cdots A_n} \to \frac{R}{A_1} \times \frac{R}{A_2} \times \cdots \times \frac{R}{A_n}$$

defined by $\varphi(\eta(\alpha)) = (\eta_1(\alpha), \eta_2(\alpha), \ldots, \eta_n(\alpha))$ is a ring isomorphism where $\eta : R \to R/A_1 \cdots A_n$ is the canonical quotient map.

Proof. The map $\theta : R \to R/A_1 \times \cdots \times R/A_n$ given by $\theta(\alpha) = (\eta_1(\alpha), \ldots, \eta_n(\alpha))$ is a ring homomorphism. It is onto by Theorem 3.24. Also $\ker \theta = A_1 \cap \cdots \cap A_n = A_1 A_2 \cdots A_n$. □

Theorem 3.26

Let A, B be integral ideals of R. Then $\exists$ an integral ideal C with $C + B = R$ such that AC is a principal ideal.

Proof. Let

$$A = P_1^{a_1} P_2^{a_2} \cdots P_n^{a_n},$$

$$B = P_1^{b_1} P_2^{b_2} \cdots P_n^{b_n},$$

with a_i, b_i integers greater than or equal to 0 and $P_1, \ldots, P_n$ distinct prime ideals. Since $P_i^{a_i+1} \subsetneq P_i^{a_i}$ we can pick an element $\alpha_i \in P_i^{a_i}$ with $\alpha_i \in P_i^{a_i+1}$ $(1 \leq i \leq n)$. Since $P_1^{a_1+1}, P_2^{a_2+1}, \ldots, P_n^{a_n+1}$ are pairwise relatively prime, by Theorem 3.24 we get an $\alpha \in R$ with

$$\alpha \equiv \alpha_i \pmod{P_i^{a_i+1}} \qquad 1 \leq i \leq n.$$

Then $\alpha \equiv \alpha_i$ (mod $P_i^{a_i}$) since $P_i^{a_i+1} \subset P_i^{a_i}$. Hence $\alpha \equiv 0$ (mod $P_i^{a_i}$), yielding $\alpha \in P_1^{a_1} \cap \cdots \cap P_n^{a_n} = P_1^{a_1} \cdots P_n^{a_n}$. From $\alpha R \subset P_1^{a_1} \cdots P_n^{a_n}$ we get $\alpha R = P_1^{a_1} \cdots P_n^{a_n} C$ for an integral ideal C. Let $C = Q_1^{\lambda_1} \cdots Q_r^{\lambda_r}$ with $\lambda_j > 0$ for each j, $Q_1, \ldots, Q_r$ distinct prime ideals of R. Since $\alpha \notin P_i^{a_i+1}$ we see immediately that $Q_1, \ldots, Q_r$ are different from any of the P_i's. Thus $\alpha R = AC$ and $B + C = R$. $\square$

Corollary 3.27

Let A be an integral ideal of R. Then $\exists$ elements p, σ in R with $A = pR + \sigma R$.

Proof. Let $0 \neq p \in A$. Set $B = pR$. From Theorem 1.43 $\exists$ an integral ideal C with $pR + C = R$ and $AC = \sigma R$ for some $\sigma \in R$. Then

$$pR + \sigma R \subset A + AC \subset A. \tag{i}$$

Also $\exists \lambda \in R$ and $x \in C$ with $1 = p\lambda + x$ since $pR + C = R$. Let $a \in A$. Then $a = ap\lambda + ax \in pR + AC$. Thus

$$A \subset pR + AC = pR + \sigma R. \tag{ii}$$

(i) and (ii) together yield $A = pR + \sigma R$. $\square$

Corollary 3.28

Let $P_1, \ldots, P_n$ be distinct prime ideals in R, $a_1, \ldots, a_n$ integers greater than or equal to 0. Then $\exists$ an $\alpha \neq 0$ in R and an integral ideal C prime to all P_i with $\alpha R = P_1^{a_1} \cdots P_n^{a_n} C$.

Proof. Take $A = P_1^{a_1} \cdots P_n^{a_n}$ and $B = P_1 \cdots P_n$ in Theorem 3.26. $\square$

Corollary 3.29

Let A, B be integral ideals. Then in R-mod we have $R/A \simeq B/AB$.

Proof. From Theorem 3.26 we see that there exists an integral ideal C of R with $A + C = R$ and $BC = pR$ with $0 \neq p \in R$. Let $\varphi: R \to (pR + AB)/AB$ be the map $\varphi(x) = px + AB$. From $pR + AB = CB + AB = (C + A)B = RB = B$ we see that $(pR + AB)/AB = B/AB$. Clearly φ is onto. Also $\varphi(x) = 0 \Leftrightarrow px \in AB$.

Thus $\varphi(x) = 0 \Rightarrow px \in AB \Rightarrow pxC \subset ABC = ApR = pA$. But $pxC \subset pA$ $\Rightarrow xC \subset A$ (since R is an integral domain). From $A + C = R$ we get $1 = a + c$ with $a \in A$ and $c \in C$. Hence $x = xa + xc$. But $xa \in A$ and

$xc \in xC \subset A$. Thus $x \in A$. Thus

$$\varphi(x) = 0 \Rightarrow x \in A. \tag{i}$$

Conversely, suppose $x \in A$. Then $\varphi(x) = px + AB$ with $p \in BC$. Thus $px \in BCA$ or $px \in AB$. Hence $\varphi(x) = 0$. In other words

$$x \in A \Rightarrow \varphi(x) = 0. \tag{ii}$$

Combining (i) and (ii) we get $\ker \varphi = A$. Thus φ induces an R isomorphism $R/A \xrightarrow[\cong]{\bar{\varphi}} B/AB$. $\square$

Definition 3.30

Two fractional ideals A, B are said to be equivalent if $\exists$ a $\gamma \in K^*$ with $A = \gamma B$.

Let $\mathcal{N} = \{A \in J | A = R\gamma \text{ for some } \gamma \in K^*\}$. $\mathcal{N}$ is a subgroup of J. In fact A, B in $\mathcal{N} \Rightarrow A = R\alpha$ and $B = R\beta$ with α, β in $K^* \Rightarrow AB^{-1} = R\alpha\beta^{-1} \in \mathcal{N}$. The quotient group $J/\mathcal{N}$ is called the *ideal class group* of the Dedekind domain R and is denoted by $Cl(R)$. If R is a PID the ideal class group of R is trivial. When R is the ring of algebraic integers in an algebraic number field K, the ideal class group of R is sometimes referred to as the class group of K. In this case it is a classical result that this class group is finite. The order of the class group of K is called the *class number* of K.

Given a fractional ideal A of a Dedekind domain R with quotient field K the map $K \otimes_R A \xrightarrow{\varphi_A} KA = K$ given by $\varphi_A(\lambda \otimes x) = \lambda x$ is a K-module isomorphism. The inverse $\theta_A: KA \to K \otimes_R A$ is given by $\theta_A(\Sigma_{i=1}^r \lambda_i x_i) = \Sigma_{i=1}^r \lambda_i \otimes_R x_i$ for $\lambda_i \in K$ and $x_i \in A$. To show that θ_A is well defined we have to prove that $\lambda_i \in K$, $x_i \in A$, and $\Sigma_{i=1}^r \lambda_i x_i = 0 \Rightarrow \Sigma_{i=1}^r \lambda_i \otimes_R x_i = 0$ in $K \otimes_R A$. We know that $\exists$ a $d \in R^*$ with $dx_i = y_i \in R$. Now

$$d\left(\sum_{i=1}^r \lambda_i \otimes_R x_i\right) = \sum_{i=1}^r \lambda_i d \otimes_R x_i$$

$$= \sum_{i=1}^r \lambda_i \otimes_R 1 \, dx_i \quad (\text{since } d \in R)$$

$$= \sum_{i=1}^r \lambda_i \, dx_i \otimes_R 1 \quad (\text{since } dx_i \in R)$$

$$= d\{(\Sigma\lambda_i x_i) \otimes_R 1\} = 0 \quad (\text{since } \Sigma\lambda_i x_i = 0).$$

Since $K \otimes_R A$ is a vector space over K and $d \in K^*$ we see that $\Sigma_{i=1}^r \lambda_i \otimes_R x_i = 0$. It is easily checked that $\theta_A \circ \varphi_A = \mathrm{Id}_{K \otimes_R A}$ and $\varphi_A \circ \theta_A = \mathrm{Id}_{KA}$. Moreover if f is any R homomorphism $A \xrightarrow{f} B$ between fractional ideals $\mathrm{Id}_K \otimes_R f : K \otimes_R A \to K \otimes_R B$ is a K homomorphism. f admits of a unique extension $g : KA = K \to KB = K$ as a K homomorphism given by $g(\sum_{i=1}^r \lambda_i a_i) = \sum_{i=1}^r \lambda_i f(a_i)$. Moreover

$$
\begin{array}{ccc}
K \otimes_R A & \xrightarrow{1 \otimes f} & K \otimes_R B \\
\downarrow{\varphi_A} & & \downarrow{\varphi_B} \\
K = KA & \xrightarrow{g} & KB = K
\end{array}
$$

is clearly commutative.

If $g(1) = \gamma \in K$, then $g(x) = xg(1) = x\gamma = \gamma x$ for all $x \in K$. In particular $f(a) = \gamma a$ for all $a \in A$. In particular it follows that $\gamma A \subset B$. Moreover, $f = 0 \Leftrightarrow \gamma = 0$. When $\gamma \in K^*$, g is an isomorphism, and f itself is a monomorphism $A \rightarrowtail B$. Also if it happens that $A \supset R$, then $f(a) = af(1) \; \forall \; a \in A$.

Lemma 3.31

Two fractional ideals A, B of R are isomorphic in R-mod $\Leftrightarrow A$ and B are equivalent fractional ideals.

Proof. Suppose $A = \gamma B$ with $\gamma \in K^*$. Then $f : B \to A$ given by $f(x) = \gamma x$ is an R isomorphism of B onto A.

Conversely, let $f : A \to B$ be an R isomorphism. If $g : KA = K \to KB = K$ is the unique extension of f as a K isomorphism (observe that $1 \otimes f : K \otimes_R A \simeq K \otimes_R B$), the comments preceding the statement of Lemma 3.31 show that $f(a) = a\gamma$ where $\gamma = g(1)$. Since f is an isomorphism, we see that $\gamma \neq 0$ in K and that $B = A\gamma$. $\square$

Theorem 3.32

Let

$$
M = I_1 \oplus \cdots \oplus I_m,
$$

$$
N = J_1 \oplus \cdots \oplus J_n,
$$

with I_i, J_j fractional ideals of R. Then in R-mod,

$$
M \simeq N \Leftrightarrow \begin{cases} m = n, & (3.33) \\ \gamma I_1 \cdots I_m = J_1 \cdots J_m & \text{for some } \gamma \in K^*. \quad (3.34) \end{cases}
$$

Proof. Suppose $M \simeq N$. Then $m = \dim_K K \otimes_R M = \dim_K K \otimes_R N = n$. This yields (3.33).

Let $I_i = A_i^{-1}$ with $A_i \in J$. For each A_i we can find a $d_i \in R^*$ with $d_i A_i \subset R$. Hence $d_i^{-1} A_i^{-1} \supset R$ or $d_i^{-1} I_i \supset R$. Similarly we can find a $c_j \in R^*$

with $c_j^{-1} J_j \supset R$. Since the ideal class of the product $I_1 \cdots I_m$ is the same as that of $d_1^{-1} I_1 \cdot d_2^{-1} I_2 \cdot d_m^{-1} I_m$ (similarly the ideal class of $J_1 \cdots J_m$ is that of $c_1^{-1} J_1 \cdots c_m^{-1} J_m$) and since $I_i \simeq d_i^{-1} I_i$ and $J_j \simeq c_j^{-1} J_j$ in R-mod for proving (3.34) we can without loss of generality assume that $I_i \supset R$ and $J_j \supset R$ for each i and for each j.

Let $\theta: I_1 \oplus \cdots \oplus I_m \simeq J_1 \oplus \cdots \oplus J_m$. Let the element 1 or R thought of as an element of I_i be denoted by 1_i. Let $p_j: J_1 \oplus \cdots \oplus J_m \to J_j$ be the projection onto the jth factor and $\theta_i = \theta | I_i$. Let $p_j \circ \theta_i(1_i) = \alpha_{ij} \in J_j$. Then from the comments preceding the statement of Lemma 3.31, we see that $p_j \circ \theta_i(x) = \alpha_{ij} x \quad \forall \ x \in I_i$. Hence $p_j \circ \theta_i(I_i) = \alpha_{ij} I_i$. Since $\theta: I_1 \oplus \cdots \oplus I_m \to J_1 \oplus \cdots \oplus J_m$ is onto we see that

$$\alpha_{1j} I_1 + \alpha_{2j} I_2 + \cdots + \alpha_{mj} I_m = J_j, \qquad 1 \le j \le m.$$

Hence for any permutation σ of $(1, 2, \ldots, m)$ we have

$$J_1 J_2 \cdots J_m \supset \alpha_{1\sigma(1)} \alpha_{2\sigma(2)} \cdots \alpha_{m\sigma(m)} I_1 \cdots I_m.$$

Hence

$$J_1 \cdots J_m \supset \delta I_1 \cdots I_m \quad \text{where } \delta = \det(\alpha_{ij}). \tag{i}$$

Let 1_j be the element 1 or R thought of as an element of J_j and let $q_i: I_1 \oplus \cdots \oplus I_m \to I_i$ be the projection onto the ith factor, $\varphi = \theta^{-1} = J_1 \oplus \cdots \oplus J_m \to I_1 \oplus \cdots \oplus I_m$, $\varphi_j = \varphi | J_j$. If $q_i \cdot \varphi_j(1_j) = \beta_{ji} \in I_i$, then the same argument as before yields

$$I_1 \cdots I_m \supset \eta J_1 \cdots J_m \quad \text{where } \eta = \det(\beta_{ij}). \tag{ii}$$

We know that

$$\varphi(\alpha_{i1}, \alpha_{i2}, \ldots, \alpha_{im}) = (0, 0, \ldots, 1_i, 0, \ldots, 0)$$

or

$$(0, 0, \ldots, 1_i, 0, \ldots, 0) = \left(\sum_{j=1}^{m} q_1 \varphi_j(\alpha_{ij}), \sum_{j=1}^{m} q_2 \varphi_j(\alpha_{ij}), \ldots, \sum_{j=1}^{m} q_m \varphi_j(\alpha_{ij}) \right)$$

or

$$\sum_{j=1}^{m} \alpha_{ij} \beta_{jk} = \begin{cases} 0 & \text{for } k \ne i, \\ 1 & \text{for } k = i. \end{cases}$$

Thus the product $(\alpha_{ij})(\beta_{ij})$ equals the identity matrix. Hence $\delta \eta = 1$. Hence (ii) can be rewritten as

$$I_1 \cdots I_m \supset \delta^{-1} J_1 \cdots J_m$$

or

$$\delta I_1 \cdots I_m \supset J_1 \cdots J_m. \tag{ii$'$}$$

Now (i) and (ii$'$) together yield

$$\delta I_1 \cdots I_m = J_1 \cdots J_m \quad \text{with } \delta \in K^*. \qquad \square$$

Conversely, suppose (3.33) and (3.34) in the statement of Theorem 3.32 are valid. We want to show that $M \simeq N$. This is an immediate consequence of the following.

Lemma 3.35

For any $m \geq 2$ and any m elements $A_1, \ldots, A_m$ in J we have

$$R^{m-1} \oplus A_1 A_2 \cdots A_m \simeq A_1 \oplus A_2 \oplus \cdots \oplus A_m \quad \text{in } R\text{-mod.}$$

Proof. First we prove this for $m = 2$. There exist α, β in R^* with αA_1^{-1} and βA_2 integral ideals. From Theorem 3.26, $\exists$ an integral ideal C with

$$\beta A_2 + C = R \quad \text{and} \quad \alpha A_1^{-1} C = pR \text{ with } 0 \neq p \in R.$$

Then $C = \alpha^{-1} p R A_1 = \gamma A_1$, say.

Thus $\gamma A_1, \beta A_2$ are fractional ideals in the equivalence classes of A_1, A_2, respectively, moreover $\gamma A_1 + \beta A_2 = C + \beta A_2 = R$. Thus without loss of generality we can assume that A_1, A_2 are *coprime integral* ideals of R. Thus there exist $\lambda_1 \in A_1$ and $\lambda_2 \in A_2$, with $\lambda_1 - \lambda_2 = 1$. Consider the map $\varphi : A_1 \oplus A_2 \to R \oplus A_1 A_2$ given by $\varphi(x_1, x_2) = (x_1 + x_2, \lambda_1 x_2 + \lambda_2 x_1)$. Clearly φ is an R-homomorphism. Given any $\lambda \in R$ and any $\alpha \in A_1 A_2$, consider

$$x_1 = \lambda_1 \lambda - \alpha \in A_1,$$

$$x_2 = \alpha - \lambda_2 \lambda \in A_2.$$

We have $\varphi(x_1, x_2) = (\lambda_1 \lambda - \lambda_2 \lambda, \lambda_1 \alpha - \lambda_1 \lambda_2 \lambda + \lambda_2 \lambda_1 \lambda - \lambda_2 \alpha) = (\lambda, \alpha)$. Hence $\varphi : A_1 \oplus A_2 \to R \oplus A_1 A_2$ is onto. Also

$$\varphi(x_1, x_2) = 0 \Rightarrow \left\{ \begin{array}{c} x_1 + x_2 = 0 \\ \lambda_2 x_1 + \lambda_1 x_2 = 0 \end{array} \right\} \Rightarrow (\lambda_1 - \lambda_2) x_1 = 0,$$

$$(\lambda_2 - \lambda_1) x_2 = 0 \Rightarrow x_1 = 0, \qquad -x_2 = 0 \Rightarrow (x_1, x_2) = (0, 0).$$

Hence $\ker \varphi = 0$. Thus $\varphi : A_1 \oplus A_2 \simeq R \oplus A_1 A_2$.

Let now $m > 2$ and assume Lemma 3.35 valid for $m - 1$ in place of m. Then

$$R^{m-2} \oplus A_2 \cdots A_m \simeq A_2 \oplus \cdots \oplus A_m.$$

Hence

$$A_1 \oplus A_2 \oplus \cdots \oplus A_m \simeq R^{m-2} \oplus (A_1 \oplus A_2 \cdots A_m)$$

$$\simeq R^{m-2} \oplus R \oplus A_1 A_2 \cdots A_m = R^{m-1} \oplus A_1 \cdots A_m. \quad \square$$

Lemma 3.36

Let R be an integral domain and I an integral ideal or R. If I is invertible, then I is projective as an R-module.

Proof. If I is invertible, then $I^{-1} = (R : I) \subset K$ satisfies $I^{-1} \cdot I = R$ where K is the quotient field of R. Hence $\exists \; \lambda_1, \ldots, \lambda_k$ in I^{-1} and $x_1, \ldots, x_k$ in I with $\sum_{i=1}^{k} \lambda_i x_i = 1$. Also $\lambda_i I \subset R$ since $\lambda_i \in I^{-1}$. Let $f_i \in \mathrm{Hom}_R(I, R)$ be given by $f_i(x) = \lambda_i x$. Observe that $\lambda_i I \subset R$. Then $\forall \; a \in I$, $\sum_{i=1}^{k} f_i(a) x_i = \sum_{i=1}^{k} \lambda_i a x_i = \sum_{i=1}^{k} \lambda_i x_i a = a$. Thus $\{ f_i; x_i \}_{i=1}^{k}$ is a projective coordinate system for I. $\quad \square$

It follows that any Dedekind domain R is hereditary. Hence from Kaplansky's result ([4], Cartan and Eilenberg, Chapter I, Theorem 5.3) any f.g. projective R-module P is isomorphic to a finite direct sum $I_1 \oplus \cdots \oplus I_n$ with I_i ideals in R.

Let $\mathrm{Cl}(R)$ denote the ideal class group of R. For any f.g. projective R-module $P \simeq I_1 \oplus \cdots \oplus I_n$ let $\langle\langle P \rangle\rangle$ be the class of $I_1 I_2 \cdots I_n$ in $\mathrm{Cl}(R)$. From Theorem 3.32 it is clear that $\langle\langle P \rangle\rangle$ depends only on the isomorphism class of P. Also if $P \oplus R^k \simeq Q \oplus R^\ell$ with P, Q f.g. projective over R we have $\langle\langle P \rangle\rangle = \langle\langle P \oplus R^k \rangle\rangle = \langle\langle Q \oplus R^\ell \rangle\rangle = \langle\langle Q \rangle\rangle$. If $P \simeq I_1 \oplus \cdots \oplus I_m$ and $Q \simeq J_1 \oplus \cdots \oplus J_n$, then $P \oplus Q \simeq I_1 \oplus \cdots \oplus I_m \oplus J_1 \oplus \cdots \oplus J_n$ and hence $\langle\langle P \oplus Q \rangle\rangle = \langle\langle P \rangle\rangle \langle\langle Q \rangle\rangle$ in $\mathscr{C}(R)$.

Theorem 3.37

The map $[P] \mapsto \langle\langle P \rangle\rangle$ gives an isomorphism $\tilde{K}_0(R) \to \mathrm{Cl}(R)$ for any Dedekind domain R.

Proof. We have only to use Theorem 3.32 to get this result. $\quad \square$

Remark 3.38

If $\omega = \exp(2\pi\sqrt{-1}/23)$, it is known that $\mathrm{Cl}(Z[\omega]) \neq \{1\}$.

Chapter Five

Dock Sang Rim's Theorem

Two good references for the material in this Chapter are [4] and [17].

1 "COMPLETE" COHOMOLOGY OF π

Throughout this chapter π will denote a finite group, $Z\pi$ the group ring of π over Z, and $I = I\pi$ the augmentation ideal. All the modules we consider will be left modules. By a π-module we thus mean a left π-module.

The element $N = \Sigma_{\sigma \in \pi}\sigma$ is called the norm element in $Z\pi$. For any $A \in Z\pi$-mod, N induces a homomorphism, also called the norm homomorphism, defined by $N(x) = N \cdot x \; \forall \; x \in A$. Observe that $\sigma N = N = N\sigma$ for any $\sigma \in \pi$. Hence $N(\sigma x) = N(x) = \sigma N(x)$ for any $x \in A$. Thus $N: A \to A$ is a π homomorphism. We know that I is generated over $Z\pi$ by elements of the form $\sigma - 1$ with $\sigma \in \pi$. From $N(\sigma - 1) = 0$ we see immediately that $N(IA) = 0$.

Definition 1.1

For any $A \in Z\pi$-mod, let $A_\pi = A/IA$ and $A^\pi = \{a \in A | \sigma \cdot a = a \; \forall \; \sigma \in \pi\}$. Then from $\sigma \cdot N = N$ we get $N(A) \subset A^\pi$. Let $_N A = \ker N: A \to A$. Then as commented already $IA \subset {_N A}$. Thus N induces maps $N^*: A_\pi \to A^\pi$ and $\bar{N}: A_\pi \to N(A)$ with $\bar{N}$ an epimorphism. If $h: N(A) \to A^\pi$ denotes the inclusion, then

$$
\begin{array}{ccc}
& N(A) & \\
\bar{N} \nearrow & & \searrow h \\
A_\pi & \xrightarrow{N^*} & A^\pi
\end{array}
$$

is a commutative diagram.

For any two π-modules A, C let $\text{Hom}(A, C) = \text{Hom}_Z(A, C)$ and $A \otimes C = A \otimes_Z C$. Then $A \otimes C$ and $\text{Hom}(A, C)$ are converted into π-modules in a natural way: $\forall\, f \in \text{Hom}(A, C)$, $a \in A$, and $\sigma \in \pi$,

$$\{\sigma f\}(a) = \sigma \cdot f(\sigma^{-1}a)$$

and $\forall\, a \in A$, $c \in C$, and $\sigma \in \pi$, $\sigma(a \otimes c) = \sigma a \otimes \sigma c$.

Remark 1.2

For any A, C in π-mod we have

$$\text{Im}\{N: \text{Hom}(A, C) \to \text{Hom}(A, C)\} \subset \text{Hom}_\pi(A, C).$$

In fact, $\text{Hom}(A, C)^\pi = \text{Hom}_\pi(A, C)$.

Proof. Let $f \in \text{Hom}_\pi(A, C)$. Then $f(\sigma a) = \sigma f(a)$ for any $a \in A$ and hence $\{\sigma f\}(a) = \sigma \cdot f(\sigma^{-1}a) = \sigma\sigma^{-1} \cdot f(a) = f(a)$. Thus $\sigma f = f$ or

$$\text{Hom}_\pi(A, C) \subset \text{Hom}(A, C)^\pi. \tag{a}$$

Conversely, let $f \in \text{Hom}(A, C)^\pi$. Then $\sigma f = f$ for every $\sigma \in \pi$ or $\sigma \cdot f(\sigma^{-1}a) = f(a) \; \forall\, a \in A$ or $f(\sigma^{-1}a) = \sigma^{-1}f(a)$ for every $\sigma \in \pi$ and $a \in A$. Hence $f \in \text{Hom}_\pi(A, C)$. Thus

$$\text{Hom}(A, C)^\pi \subset \text{Hom}_\pi(A, C). \tag{b}$$

We know that $N(\text{Hom}(A, C)) \subset \text{Hom}(A, C)^\pi$. This proves Remark 1.2. $\quad\square$

Remark 1.3

If $f \in \text{Hom}_\pi(A, C)$, then $N(f) = |\pi|f$ where $|\pi|$ equals the order of the group π. In fact

$$\{N(f)\}(a) = \sum_{\sigma \in \pi} \{\sigma f\}(a) = \sum_{\sigma \in \pi} \sigma \cdot f(\sigma^{-1}a) = \sum_{\sigma \in \pi} \sigma \cdot \sigma^{-1}f(a)$$

$$= |\pi|f(a) \qquad \forall\, a \in A.$$

Remark 1.4

Let A, B, C, D, be any four π-modules, $f \in \text{Hom}_\pi(A, B)$, $h \in \text{Hom}_\pi(C, D)$, and $g \in \text{Hom}(B, C)$.

Then in $\text{Hom}(A, D)$ we have $N(hgf) = h \circ Ng \circ f$.

Proof. Straightforward. $\square$

Let $A \in \pi\text{-mod}$.

Definition 1.5

The usual homology and cohomology groups of π with coefficients in A are defined by $H_n(\pi, A) = \text{tor}_n^\pi(Z, A)$ for $n \geq 0$, where Z is regarded as a trivial *right* π-module. $H^n(\pi, A) = \text{Ext}_\pi^n(Z, A)$ for $n \geq 0$, where Z is regarded as a trivial left π-module.

One knows that $H_0(\pi, A) = A_\pi$ and $H^0(\pi, A) = A^\pi$.

Definition 1.6

The "complete derived series" of π with coefficients in A or the "complete cohomology" of π with coefficients in A is the graded abelian group $\hat{H}(\pi, A) = \oplus_{n \in Z} \hat{H}^n(\pi, A)$ where

$$\hat{H}^n(\pi, A) = H^n(\pi, A) \quad \text{for } n \geq 1,$$

$$\hat{H}^0(\pi, A) = \text{coker}\big\{ H_0(\pi, A) = A_\pi \xrightarrow{N*} A^\pi = H^0(\pi, A)\big\} = A^\pi/N(A),$$

$$\hat{H}^{-1}(\pi, A) = \ker\big\{ H_0(\pi, A) = A_\pi \xrightarrow{N*} A^\pi = H^0(\pi, A)\big\} = {}_NA/IA,$$

$$\hat{H}^{-n-1}(\pi, A) = H_n(\pi, A) \quad \text{for } n \geq 1.$$

Lemma 1.7

Let $0 \to A' \xrightarrow{i} A \xrightarrow{\eta} A'' \to 0$ be any exact sequence of π-modules. Then the boundary map $\partial\colon H_1(\pi, A'') \to H_0(\pi, A') = A'/IA' = A'_\pi$ satisfies $\partial(H_1(\pi, A'')) \subset N^{A'}/IA'$.

Proof. We know that $H_1(\pi, A'') \xrightarrow{\partial} H_0(\pi, A') \xrightarrow{i_*} H_0(\pi, A) \xrightarrow{\eta_*} H_0(\pi, A'') \to 0$ is exact.

Thus $H_1(\pi, A'') \overset{\partial}{\to} A'/IA' \overset{i_*}{\longrightarrow} A/IA \overset{\eta_*}{\longrightarrow} A''/IA'' \to 0$ is exact. Hence

$$\partial\left(H_1(\pi, A'')\right) = \ker i_*: A'/IA' \to A/IA$$

$$= \frac{A' \cap IA}{IA'} \quad \text{(identifying } A' \text{ with a submodule of } A \text{ using } i\text{)}.$$

But we know $IA \subset \ker N: A \to A$ and $N: A' \to A'$ is obtained from restricting N to A' (rather

$$
\begin{array}{ccc}
A' & \overset{N}{\longrightarrow} & A' \\
i \downarrow & & \downarrow i \\
A & \overset{N}{\longrightarrow} & A
\end{array}
$$

is commutative). Thus $A' \cap IA \subset {}_N A'$. Hence $\partial(H_1(\pi, A'')) \subset {}_N A'/IA'$. Observe that ${}_N A'/IA' = \hat{H}^{-1}(\pi, A')$ and $H_1(\pi, A'') = \hat{H}^{-2}(\pi, A'')$ both by definition. Thus Lemma 1.7 can be restated as

$$\partial\left(\hat{H}^{-2}(\pi, A'')\right) \subset \hat{H}^{-1}(\pi, A'). \qquad \square$$

Lemma 1.8

Let $0 \to A' \overset{i}{\to} A \overset{\eta}{\to} A'' \to 0$ be an exact sequence of π-modules. Let $\partial: H^0(\pi, A'') = A''^\pi \to H^1(\pi, A')$ be the boundary map in the usual cohomology exact sequence of $0 \to A' \to A \to A'' \to 0$. Then $\partial(N(A'')) = 0$ and hence ∂ induces a map $\bar{\partial}: A''^\pi/N(A'') \to H^1(\pi, A')$.

     Proof. We know that $\to H^0(\pi, A) \overset{\eta_*}{\longrightarrow} H^0(\pi, A'') \overset{\partial}{\to} H^1(\pi, A') \overset{i_*}{\longrightarrow} H^1(\pi, A) \to \cdots$ is exact. Thus

$$A^\pi \overset{\eta_* = \eta/A^\pi}{\longrightarrow} A''^\pi \overset{\partial}{\longrightarrow} H^1(\pi, A') \quad \text{is exact.}$$

We also know that $N(A) \subset A^\pi$. Given $a'' \in A''$ we have some $a \in A$ with $\eta(a) = a''$. Hence $\eta(Na) = Na''$. Thus $\eta|N(A): N(A) \to N(A'')$ is onto. It follows that $\eta(N(A)) = N(A'') \subset \ker \partial: A''^\pi \to H^1(\pi, A')$. $\square$

We will need Diagram 1.9 for our proof of Theorem 1.10. In this diagram the maps indicated by μ are inclusions and the maps indicated by ν's are quotient maps (epimorphisms).

$$
\begin{array}{ccccccc}
& 0 & & 0 & & 0 & \\
& \downarrow & & \downarrow & & \downarrow & \\
H_1(\pi, A'') \xrightarrow{\partial} & N^{A'}/IA' & \xrightarrow{i_*} & N^A/IA & \xrightarrow{\eta_*} & N^{A''}/IA'' & \\
\| & \downarrow \mu & & \downarrow \mu & & \downarrow \mu & \\
H_1(\pi, A'') \xrightarrow{\partial} & A'/IA' & \xrightarrow{i_*} & A/IA & \xrightarrow{\eta_*} & A''/IA'' & \longrightarrow 0 \qquad (1) \\
& \downarrow N^* & & \downarrow N^* & & \downarrow N^* & \\
0 \longrightarrow & A'^\pi & \xrightarrow{i_*} & A^\pi & \xrightarrow{\eta_*} & A''^\pi & \xrightarrow{\partial} H^1(\pi, A') \quad (2) \\
& \downarrow \nu & & \downarrow \nu & & \downarrow \nu & \| \\
& A'^\pi/N(A') \xrightarrow{i_*} & A^\pi/N(A) & \xrightarrow{\eta_*} & A''^\pi/N(A'') & \xrightarrow{\bar\partial} & H^1(\pi, A') \\
& \downarrow & & \downarrow & & \downarrow & \\
& 0 & & 0 & & 0 & \\
& (V_1) & & (V_2) & & (V_3) &
\end{array}
$$

Diagram 1.9.

In this diagram the sequences marked $(V_1), (V_2), (V_3)$ and $(1), (2)$ are exact.

Theorem 1.10

Let $0 \to A' \xrightarrow{i} A \xrightarrow{\eta} A'' \to 0$ be an exact sequence of π-modules. Then there exists an exact sequence

$$
\cdots \longrightarrow \hat{H}^n(\pi, A') \xrightarrow{i_*} \hat{H}^n(\pi, A) \xrightarrow{\eta_*} \hat{H}^n(\pi, A'') \xrightarrow{\partial} \hat{H}^{n+1}(\pi, A') \longrightarrow \cdots
$$

where n varies over Z.

Proof. For $n \geq 1$, it is the same as the part

$$
H^1(\pi, A') \xrightarrow{i_*} H^1(\pi, A) \xrightarrow{\eta_*} H^1(\pi, A'') \xrightarrow{\partial} H^2(\pi, A') \longrightarrow \cdots
$$

of the cohomology exact sequence. For $n \leq -2$ it is the same as the part

$$
\begin{array}{cccccccccc}
\cdots \to H_2(\pi, A') & \xrightarrow{i_*} & H_2(\pi, A) & \xrightarrow{\eta_*} & H_2(\pi, A'') & \xrightarrow{\partial} & H_2(\pi, A') & \xrightarrow{i_*} & H_1(\pi, A) & \xrightarrow{\eta_*} & H_1(\pi, A'') \\
\| & & \| & & \| & & \| & & \| & & \| \\
\cdots \to \hat{H}^{-3}(\pi, A') & \longrightarrow & \hat{H}^{-3}(\pi, A) & \longrightarrow & \hat{H}^{-3}(\pi, A'') & \longrightarrow & \hat{H}^{-2}(\pi, A') & \longrightarrow & \hat{H}^{-2}(\pi, A) & \longrightarrow & \hat{H}^{-2}(\pi, A'')
\end{array}
$$

of the homology exact sequence.

In Diagram 1.9 we know that sequences $(1), (2), (V_1), (V_2), (V_3)$ are exact. It can easily be checked that

$$H_1(\pi, A'') \xrightarrow{\partial} {}_N A'/IA' \xrightarrow{i_*} {}_N A/IA \xrightarrow{\eta_*} {}_N A''/IA''$$

$$A'^{\pi}/N(A') \xrightarrow{i_*} A^{\pi}(N(A) \xrightarrow{\eta_*} A''^{\pi}/N(A'') \xrightarrow{\bar{\partial}} H^1(\pi, A')$$

are exact.

Let $x'' \in {}_N A''/IA''$. There exists an $a \in A/IA$ with $\eta_*(a) = \mu(x'')$. Then $\eta_* N^*(a) = N^* \eta_*(a) = N^* \mu(x'') = 0$.

Hence $\exists$ an element $c' \in A'^{\pi}$ with $i_*(c') = N^*(a)$. Define $\partial: N^{A''}/IA \to A'^{\pi}/N(A')$ by $\partial x'' = \nu(c')$ where $c' \in A'^{\pi}$ and $a \in A/IA$ are such that $i_*(c') = N^*(a)$ and $\eta_*(a) = \mu(x'')$.

Suppose $b \in A/IA$ is another element with $\eta_*(b) = \mu(x'')$. Then $\eta_*(b - a) = 0$. Hence $b - a = i_*(u')$ with $u' \in A'/IA'$. If $\lambda' = c' + N^*(u')$ in A'^{π} we have

$$i_*(\lambda') = i_*(c') + i_* N^*(u')$$

$$= N^*(a) + N^* i_*(u') = N^*(a) + N^*(b - a) = N^*(b).$$

To show that $\partial: {}_N A''/IA'' \to A'^{\pi}/N(A')$ is well defined, we have to check that $\nu(c') = \nu(\lambda')$. But $\lambda' = c' + N^*(u')$. Hence $\nu(\lambda') = \nu(c') + \nu N^*(u') = \nu(c')$ since $\nu N^*(u') = 0$. It is now easy to check that $\partial: {}_N A''/IA'' \to A'\pi/N(A')$ is a π homomorphism.

Standard diagram chasing establishes the exactness of

$$\xrightarrow{\eta_*} H_1(\pi, A'') \longrightarrow {}_N A'/IA' \xrightarrow{i_*} {}_N A/IA \xrightarrow{\eta_*} {}_N A''/IA''$$

$$\xrightarrow{\partial} A'^{\pi}/N(A') \xrightarrow{i_*} A^{\pi}/N(A).$$

This completes the proof of Theorem 1.10. $\square$

2   WEAKLY PROJECTIVE AND WEAKLY INJECTIVE MODULES

Let K_1, K_2, and K be commutative rings and $T: K_1\text{-mod} \times K_2\text{-mod} \to K\text{-mod}$ an additive functor. Let $A_1 \in K_1\pi\text{-mod}$ and $A_2 \in K_2\pi\text{-mod}$. Then $T(A_1, A_2)$ acquires in a natural way a $K\pi$-module structure. If T is covariant in both the variables, for any $\sigma \in \pi$ the action of σ on $T(A_1, A_2)$ is given by $\sigma \cdot x = T(\sigma, \sigma)(x) \forall x \in T(A_1, A_2)$. If T is contravariant in one of the variables, say in A_1, and covariant in the other variable A_2, then $\sigma \cdot x = T(\sigma^{-1}, \sigma)(x) \forall$

$x \in T(A_1, A_2)$. If T is contravariant in both the variables, then $\sigma \cdot x = T(\sigma^{-1}, \sigma^{-1})(x)$. For instance, if A, B are in $K\pi$-mod, $A \otimes_K B$ and $\mathrm{Hom}_K(A, B)$ are regarded as $K\pi$-modules by

$$\left. \begin{array}{l} \sigma \cdot (a \otimes b) = \sigma a \otimes \sigma b \\ \{\sigma f\{(a) = a \cdot f(\sigma^{-1}a)\} \end{array} \right| \quad \begin{array}{l} \text{for any } a \in A,\, b \in B,\, f \in \mathrm{Hom}_K(A, B), \\ \text{and } \sigma \in \pi. \end{array}$$

For any $A \in K\pi$-mod, the preceding considerations yield natural $K\pi$-module structures on

$$K\pi \otimes_K A \quad \text{and} \quad \mathrm{Hom}_K(K\pi, A)$$

given by $\sigma(\lambda \otimes a) = \sigma\lambda \otimes \sigma a$:

$$\{\sigma f\}(\lambda) = \sigma \cdot f(\sigma^{-1}\lambda) \quad \text{for any } \lambda \in K\pi,\, a \in A, \text{ and } \sigma \in \pi. \quad (2.1)$$

We will denote $K\pi \otimes_K A$ and $\mathrm{Hom}_K(K\pi, A)$ with the preceding $K\pi$-module structures by $(K\pi \otimes_K A)_1$ and $(\mathrm{Hom}_K(K\pi, A))_1$.

The modules (K-modules) $K\pi \otimes_K A$ and $\mathrm{Hom}_K(K\pi, A)$ can also be regarded as $K\pi$-modules taking only the π operations of $K\pi$ into consideration and neglecting the π operations on A. We will use the left π operation on $K\pi$ to convert $K\pi \otimes_K A$ into a left $K\pi$-module and right π operation on $K\pi$ to convert $\mathrm{Hom}_K(K\pi, A)$ into a left $K\pi$-module, namely

$$\left. \begin{array}{l} \sigma(\lambda \otimes a) = \sigma\lambda \otimes a \\ (\sigma f)(\lambda) = f(\lambda\sigma) \end{array} \right| \quad \begin{array}{l} \text{for any } \sigma \in \pi,\, \lambda \in K\pi,\, a \in A, \\ \text{and } f \in \mathrm{Hom}_K(K\pi, A). \end{array} \quad (2.2)$$

The $K\pi$-modules $K\pi \otimes_K A$ and $\mathrm{Hom}_K(K\pi, A)$ with π operations given by (2.2) will be denoted by

$$\left(K\pi \underset{K}{\otimes} A \right)_{(2)} \quad \text{and} \quad \mathrm{Hom}_K(K\pi, A)_{(2)}.$$

The map $\varphi: (K\pi \otimes_K A)_{(2)} \to (K\pi \otimes_K A)_{(1)}$ defined by

$$\cdots \varphi(\lambda \otimes a) = \lambda \otimes \lambda a \quad \forall\, \lambda \in K\pi,\, a \in A, \quad (2.3)$$

is a $K\pi$ isomorphism, with $\varphi^{-1}: (K\pi \otimes_K A)_{(1)} \to (K\pi \otimes_K A)_{(2)}$ given by

$$\cdots \varphi^{-1}(\sigma \otimes a) = \sigma \otimes \sigma^{-1}a \quad \forall\, \sigma \in \pi,\, a \in A \quad (2.4)$$

[or if $\lambda \to \bar{\lambda}$ denotes the K map $K\pi \to K\pi$ arising from $\sigma \to \sigma^{-1}$ of π, then $\varphi^{-1}(\lambda \otimes a) = \lambda \otimes \bar{\lambda}a \;\; \forall \;\; \lambda \in K\pi, \;\; a \in A$]. Similarly, the map $\mathrm{Hom}_K(K\pi, A)_{(2)} \overset{\Psi}{\to} \mathrm{Hom}_K(K\pi, A)_{(1)}$ defined by

$$\cdots \{ \Psi f \}(\sigma) = \sigma \cdot f(\sigma^{-1}) \tag{2.5}$$

is a $K\pi$ isomorphism. In fact for any $\tau \in \pi$,

$$\Psi((\tau f)_{(2)})(\sigma) = \sigma \cdot (\tau f)_{(2)}(\sigma^{-1}) = \sigma \cdot f(\sigma^{-1}\tau),$$

$$(\tau \Psi(f))_{(1)}(\sigma) = \tau \cdot \Psi(f)(\tau^{-1}\sigma)$$

$$= \tau \cdot \tau^{-1}\sigma f(\sigma^{-1}\tau) = \sigma \cdot f(\sigma^{-1}\tau).$$

Thus $\Psi((\tau f)_{(2)}) = (\tau \Psi(f))_{(1)}$. Thus $\Psi: \mathrm{Hom}_K(K\pi, A)_{(2)} \to \mathrm{Hom}_K(K\pi, A)_{(1)}$ is a $K\pi$ homomorphism. The inverse of Ψ is easily seen to be

$$\cdots \{ \Psi^{-1}(g) \}(\sigma) = \sigma \cdot g(\sigma^{-1}) \quad \forall \sigma \in \pi. \tag{2.6}$$

In fact

$$\{ \Psi(\Psi^{-1}(g)) \}(\sigma) = \sigma \cdot \Psi^{-1}(g)(\sigma^{-1}) \quad \text{by the definition of } \Psi$$

$$= \sigma \cdot \sigma^{-1}g((\sigma^{-1})^{-1}) = g(\sigma) \quad \forall g \in \mathrm{Hom}_K(K\pi, A)_{(1)}.$$

Similarly $\Psi^{-1}\Psi(f) = f \;\; \forall f \in \mathrm{Hom}_K(K\pi, A)_{(2)}$.

Let $A \in K\pi$-mod. Then the map $g: (K\pi \otimes_K A)_{(2)} \to A$ given by $g(\lambda \otimes a) = \lambda a$ is a $K\pi$ epimorphism. If A is projective in $K\pi$-mod, then $\exists$ a splitting for g. We now introduce the following.

Definition 2.7

A $K\pi$-module A is said to be *weakly projective* if $0 \to \ker g \to (K\pi \otimes_K A)_{(2)} \overset{g}{\to} A \to 0$ is a split exact sequence in $K\pi$-mod where g is the map given by $g(\lambda \otimes a) = \lambda a \;\; \forall a \in A$.

$\exists$ a natural injection $h: A \to \mathrm{Hom}_K(K\pi, A)_{(2)}$ defined by $\{ h(a) \}(\lambda) = \lambda a \;\; \forall \lambda \in K\pi$. In fact for any $\lambda \in \pi$,

$$\{ h(\sigma a) \}(\lambda) = \lambda(\sigma a) = \lambda \sigma \cdot a,$$

$$(\sigma h(a))_{(2)}(\lambda) = H(a)(\lambda \sigma) = \lambda \sigma \cdot a.$$

Thus h is a $K\pi$ map. Clearly $h(a) = 0 \Rightarrow 1 \cdot a = 0 \Rightarrow a = 0$ (taking $\lambda = 1 \in K\pi$).

Definition 2.8

A $K\pi$-module A is said to be *weakly injective* if

$$0 \to A \xrightarrow{h} \mathrm{Hom}_K(K\pi, A)_{(2)} \to \mathrm{coker}\, h \to 0$$

is exact in $K\pi$-mod.

Theorem 2.9

The following conditions on a $K\pi$-module A are equivalent:

   (i) $\exists$ *an* $f \in \mathrm{Hom}_K(A, A)$ *with* $Nf = \mathrm{Id}_A$.
   (ii) *A is weakly projective.*
   (iii) *A is weakly injective.*

Proof. Let $g: (K\pi \otimes_K A)_{(2)} \to A$ be the map $g(\lambda \otimes a) = \lambda a$.
Then by definition, A is weakly projective $\Leftrightarrow \exists$ a splitting for g. Since the map $(K\pi \otimes_K A)_{(1)} \xrightarrow{\varphi^{-1}} (K\pi \otimes_K A)_{(2)}$ carrying $\sigma \otimes a$ to $\sigma \otimes \sigma^{-1}a$ is a π isomorphism, we see that A is weakly projective $\Leftrightarrow g': (K\pi \otimes_K A)_{(1)} \to A$ given by $\varphi^{-1}g = g'$ admits of a splitting in $K\pi$-mod. Now $g'(\sigma \otimes a) = g(\sigma \otimes \sigma^{-1}a) = a \; \forall \; \sigma \in \pi$ and $a \in A$. With these comments we take up the proof of Theorem 2.9.
   (ii) $\Rightarrow$ (i). Assume A weakly projective. Then $\exists$ a $K\pi$ homomorphism $h: A \to (K\pi \otimes_K A)_{(1)}$ with $g'h = \mathrm{Id}_A$.
   Since the elements $\sigma \in \pi$ form a K basis for $K\pi$, the module $K\pi \otimes_K A$ as a K-module is the direct sum $\oplus_{\sigma \in \pi} K\sigma \otimes_K A$. Hence for any $a \in A$, we can write $h(a) = \Sigma_{\sigma \in \pi} \sigma \otimes \alpha(\sigma, a)$ with $\alpha(\sigma, a) \in A$, this expression being unique. Moreover for any $\sigma \in \pi$, $a \mapsto \alpha(\sigma, a)$ is a K homomorphism $A \to A$. Let $\tau \in \pi$. Then

$$h(\tau a) = \sum_{\sigma \in \pi} \sigma \otimes \alpha(\sigma, \tau a)$$

and

$$\tau \cdot h(a) = \sum_{\sigma \in \pi} \tau\sigma \otimes \alpha(\sigma, a)$$

$$= \sum_{\sigma' \in \pi} \sigma' \otimes \tau\alpha(\tau^{-1}\sigma', a), \qquad \text{where } \tau\sigma = \sigma',$$

or

$$\tau \cdot h(a) = \sum_{\sigma \in \pi} \sigma \otimes \tau\alpha(\tau^{-1}\sigma, a).$$

Hence

$$\boxed{h(\tau a) = \tau \cdot h(a) \Leftrightarrow \alpha(\sigma, \tau a) = \tau\alpha(\tau^{-1}\sigma, a)} \tag{i}$$

Let now $\beta: \pi \times A \to A$ be any function satisfying

$$\beta(\sigma, \tau a) = \tau\beta(\tau^{-1}\sigma, a) \tag{$*$}$$

for any σ, τ in π and $a \in A$. In $(*)$ take $a = \sigma^{-1}b$ and $\tau = \sigma$ with $b \in A$. Then as a particular case of $(*)$ we get

$$\beta(\sigma, b) = \sigma\beta(1, \sigma^{-1}b). \tag{$**$}$$

Conversely, suppose $\beta: \pi \times A \to A$ satisfies $(**)$. Then we claim that β satisfies $(*)$.

In fact

$$\tau \cdot \beta(\tau^{-1}\sigma, a) = \tau \cdot \left\{ \tau^{-1}\sigma \cdot \beta(1, \sigma^{-1}\tau a) \right\} \quad \text{from } (**)$$

$$= \sigma\beta(1, \sigma^{-1}\tau a)$$

$$= \sigma\beta(1, \sigma^{-1}b) \quad \text{where } b = \tau a$$

$$= \beta(\sigma, b) \quad \text{from } (**)$$

$$= \beta(\sigma, \tau a).$$

Thus $\beta(\sigma, \tau a) = \tau \cdot \beta(\tau^{-1}\sigma, a)$ for all σ, τ in π and $a \in A$. Now,

$$a = g'h(a) \quad \text{for any } a \in A$$

$$= g'\left(\sum_{\sigma \in \pi} \sigma \otimes \alpha(\sigma, a) \right)$$

$$= \sum_{\sigma \in \pi} \alpha(\sigma, a).$$

Since h is a $K\pi$ homomorphism, from (i) and the implication $(*) \Rightarrow (**)$ we see that $\alpha(\sigma, a) = \sigma \cdot \alpha(1, \sigma^{-1})$. Hence

$$a = \sum_{\sigma \in \pi} \sigma \cdot \alpha(1, \sigma^{-1}a). \tag{ii}$$

Let $f \in \mathrm{Hom}_K(A, A)$ be given by $f(a) = \alpha(1, a)$. Then $Nf(a) = \sum_{\sigma \in \pi} \sigma \cdot f(\sigma^{-1}a) = \sum_{\sigma \in \pi} \sigma \cdot \alpha(1, \sigma^{-1}a) = a$ by (ii). Thus $Nf = \mathrm{Id}_A$. This completes the proof of (ii) $\Rightarrow$ (i).

(i) $\Rightarrow$ (ii). Let $f \in \mathrm{Hom}_K(A, A)$ satisfy $Nf = \mathrm{Id}_A$. Define $\alpha \colon \pi \times A \to A$ by $\alpha(\sigma, a) = \sigma f(\sigma^{-1}a)$ and $h \colon A \to (K\pi \otimes_K A)_{(1)}$ by $h(x) = \sum_{\sigma \in \pi} \sigma \otimes \alpha(\sigma, x)$ for any $x \in A$. Then $\alpha(1, \sigma^{-1}b) = f(\sigma^{-1}b)$ and $\alpha(\sigma, b) = \sigma f(\sigma^{-1}b) = \sigma\alpha(1, \sigma^{-1}b)$. Thus α satisfies $(**)$ and hence $(*)$ and hence by (i) h is a $K\pi$ homomorphism. Also $g'h(a) = g'(\sum_{\sigma \in \pi} \sigma \otimes \alpha(\sigma, a)) = \sum_{\sigma \in \pi} \alpha(\sigma, a) = \sum_{\sigma \in \pi} \sigma \cdot f(\sigma^{-1}a) = Nf(a) = a$. Thus $h \colon A \to (K\pi \otimes_K A)_{(1)}$ is a splitting of g' and hence A is weakly projective.

(i) $\Rightarrow$ (iii). Before taking up the proof of this, we first observe that A is weakly injective $\Leftrightarrow \exists$ a $K\pi$ homomorphism $\mu \colon \mathrm{Hom}_K(K\pi, A)_{(1)} \to A$ such that $\mu \circ \Psi \circ h = \mathrm{Id}_A$, where $h \colon A \to \mathrm{Hom}_K(K\pi, A)_{(2)}$ is given by $\{h(a)\}(\lambda) = \lambda a \; \forall \, \lambda \in K\pi$.

Let $f \in \mathrm{Hom}_K(A, A)$ satisfy $Nf = \mathrm{Id}_A$. Define $\mu \colon \mathrm{Hom}_K(K\pi, A)_{(1)} \to A$ by $\mu(g) = \sum_{\sigma \in \pi} \sigma \cdot f(\sigma^{-1} \cdot g(\sigma))$. Then for any $\tau \in \pi$, $(\tau g)(\sigma) = \tau \cdot g(\tau^{-1}\sigma)$. Hence $\mu(\tau g) = \sum_{\sigma \in \pi} \sigma \cdot f(\sigma^{-1} \cdot \tau \cdot g(\tau^{-1}\sigma))$. Write $\sigma' = \tau^{-1}\sigma$. Then

$$\mu(\tau g) = \sum_{\sigma' \in \pi} \tau \cdot \sigma' \cdot f(\sigma'^{-1} \cdot g(\sigma'))$$

$$= \tau \cdot \sum_{\sigma' \in \pi} \sigma' \cdot f(\sigma'^{-1} \cdot g(\sigma')) = \tau \cdot \mu(g).$$

Thus $\mu \colon \mathrm{Hom}_K(K\pi, A)_{(1)} \to A$ is a $K\pi$ homomorphism. Moreover

$$\mu \circ \Psi \circ h(a) = \sum_{\sigma \in \pi} \sigma \cdot f(\sigma^{-1} \cdot \{\Psi \circ h(a)\}(\sigma))$$

$$= \sum_{\sigma \in \pi} \sigma \cdot f(\sigma^{-1} \cdot \sigma \cdot h(a)(\sigma^{-1}))$$

$$= \sum_{\sigma \in \pi} \sigma \cdot f(h(a)(\sigma^{-1}))$$

$$= \sum_{\sigma \in \pi} \sigma \cdot f(\sigma^{-1}a))$$

$$= Nf(a) = a.$$

Thus $\mu \circ (\Psi \circ h) = \mathrm{Id}_A$ and hence A is weakly injective.

(iii) $\Rightarrow$ (i). Let μ: $\mathrm{Hom}_K(K\pi, A)_{(1)} \to A$ satisfy $\mu \circ \Psi \circ h = \mathrm{Id}_A$, where h: $A \to \mathrm{Hom}_K(K\pi, A)_{(2)}$ is the map $h(a)(\lambda) = \lambda a$ $\forall$ $\lambda \in K\pi$, $a \in A$. For any $a \in A$, let $\theta_a \in \mathrm{Hom}_K(K\pi, A)$ be given by $\theta_a(1) = a$ and $\theta_a(\sigma) = 0$ for $\sigma \neq 1$ in π. Let $f \in \mathrm{Hom}_K(A, A)$ be given by $f(a) = \mu(\theta_a)$. Then

$$Nf(a) = \sum_{\sigma \in \pi} \sigma \cdot f(\sigma^{-1}a)$$

$$= \sum_{\sigma \in \pi} \sigma \cdot \mu(\theta_{\sigma^{-1}a})$$

$$= \sum_{\sigma \in \pi} \mu(\sigma\theta_{\sigma^{-1}a}) \quad \text{since } \mu \text{ is a } K\pi \text{ homomorphism}$$

$$= \mu\left(\sum_{\sigma \in \pi} \sigma\theta_{\sigma^{-1}a}\right). \tag{iii}$$

Consider the elements $\Psi \circ h(a)$ and $\sum_{\sigma \in \pi}\sigma\theta_{\sigma^{-1}a}$ of $\mathrm{Hom}_K(K\pi, A)$. We have

$$\{\Psi \circ h(a)\}(\sigma) = \sigma \cdot h(a)(\sigma^{-1}) \quad \text{from (2.6)}$$

$$= \sigma \cdot \sigma^{-1}a \quad \text{definition of } h(a)$$

$$= a.$$

Thus $\{\Psi \circ h(a)\}(\sigma) = a$ $\forall$ $\sigma \in \pi$. Also

$$\{\sigma\theta_{\sigma^{-1}a}\}(\tau) = \sigma \cdot \theta_{\sigma^{-1}a}(\sigma^{-1}\tau) \quad \text{by definition}$$

$$= \begin{cases} \sigma \cdot 0, & \text{if } \sigma^{-1}\tau \neq 1, \\ \sigma \cdot \sigma^{-1}a, & \text{if } \sigma^{-1}\tau = 1, \end{cases}$$

$$= \begin{cases} 0, & \text{if } \tau \neq \sigma, \\ a, & \text{if } \tau = \sigma. \end{cases}$$

Hence $(\sum_{\sigma \in \pi}\sigma\theta_{\sigma^{-1}a})(\tau) = a$ $\forall$ $\tau \in \pi$. Thus $\Psi \circ h(a) = \sum_{\sigma \in \pi}\sigma\theta_{\sigma^{-1}a}$ in $\mathrm{Hom}_K(K\pi, A)_{(1)}$. Hence

$$Nf(a) = \mu\left(\sum_{\sigma \in \pi} \sigma\theta_{\sigma^{-1}a}\right) \quad \text{by (iii)}$$

$$= \mu(\Psi \circ h(a))$$

$$= a.$$

This completes the proof of Theorem 2.9. $\square$

Remark 2.10

Let $A \in K\pi$-mod be weakly projective. Then from Theorem 2.9 we see that $\exists$ an $f \in \mathrm{Hom}_K(A, A)$ with $\sum_{\sigma \in P} \sigma \cdot f(\sigma^{-1} a) = a \ \forall \ a \in A$. Since any such f is automatically in $\mathrm{Hom}_Z(A, A)$, it follows from Theorem 2.9 for the case of $Z\pi$ that any $A \in K\pi$-mod that is weakly projective automatically satisfies the following conditions:

(i) A as a $Z\pi$-module is weakly projective.

(ii) A as a $Z\pi$-module is weakly injective.

Proposition 2.11

Let $T: K_1$-mod $\times K_2$-mod $\to K$-mod be an additive functor and $A_i \in K_i\pi$-mod, $i = 1, 2$. If one of A_1 or A_2 is weakly projective in $K_1\pi$-mod (or $K_2\pi$-mod), then $T(A_1, A_2)$ is weakly projective in $K\pi$-mod.

Proof. We will give the proof in the case when T is covariant in both the variables and A_1 is weakly projective in $K_1\pi$-mod.

From Theorem 2.9 we see that $\exists$ an $f \in \mathrm{Hom}_K(A_1, A_1)$ with $Nf = \mathrm{Id}_{A_1}$. Let $\theta = T(f, \mathrm{Id}_{A_2}) \in \mathrm{Hom}_K(T(A_1, A_2), T(A_1, A_2))$. For any $x \in T(A_1, A_2)$ and $\sigma \in \pi$ we have $\sigma \cdot x = T(\sigma, \sigma)(x)$. Hence $N\theta(x) = \sum_{\sigma \in \pi} \sigma \cdot \theta(\sigma^{-1} \cdot x)$. Equivalently,

$$N\theta(x) = \sum_{\sigma \in \pi} \sigma \cdot T(f, \mathrm{Id}_{A_2}) T(\sigma^{-1}, \sigma^{-1})(x)$$

$$= \sum_{\sigma \in \pi} T(\sigma, \sigma) \circ T(f \cdot \mathrm{Id}_{A_2}) \cdot T(\sigma^{-1}, \sigma^{-1})(x) \qquad \forall \ x \in T(A_1, A_2).$$

Equivalently,

$$N\theta = \sum_{\sigma \in \pi} T(\sigma, \sigma) \cdot T(f, \mathrm{Id}_{A_2}) \cdot T(\sigma^{-1}, \sigma^{-1})$$

$$= \sum_{\sigma \in \pi} T\left(\sigma \cdot f(\sigma^{-1} \cdots), \mathrm{Id}_{A_2}\right) = T\left(\sum_{\sigma \in \pi} \sigma \cdot f(\sigma^{-1} \cdots), \mathrm{Id}_{A_2}\right)$$

$$= T\left(\mathrm{Id}_{A_1}, \mathrm{Id}_{A_2}\right) \quad \text{since} \ \sum_{\sigma \in \pi} \sigma \cdot f(\sigma^{-1} \cdots) = Nf = \mathrm{Id}_{A_1}$$

$$= \mathrm{Id}_{T(A_1, A_2)}.$$

Hence $T(A_1, A_2)$ is $K\pi$ weakly projective. $\square$

Corollary 2.12

Let A, B in $K\pi$-mod *be weakly projective. Then* $A \otimes_K B$ *and* $\text{Hom}_K(A, B)$ *are both weakly projection in* $K\pi$-mod.

Proposition 2.13

Let $A \in Z\pi$-mod *be weakly projective. Then in the diagram*

$$
\begin{array}{ccc}
 & N(A) & \\
\bar{N} \nearrow & & \searrow h \\
A_\pi = A/IA & \xrightarrow{\ N^*\ } & A^\pi
\end{array}
$$

all the maps are isomorphisms. In particular $\hat{H}^0(\pi, A) = A^\pi/N(A) = 0$ *and* $\hat{H}^{-1}(\pi, A) = {}_N A/IA = 0$.

Proof. Since A is weakly projective $\exists$ an $f \in \text{Hom}_Z(A, A)$ with $Nf = \text{Id}_A$. Let $a \in \ker N\colon A \to A$, i.e., $Na = 0$. We have $a = \sum_{\sigma \in \pi} \sigma \cdot f(\sigma^{-1}a)$. Also $Na = \sum_{\sigma \in \pi}\sigma^{-1}a = 0$. Hence $\sum_{\sigma \in \pi} f(\sigma^{-1}a) = 0$. Hence

$$
\begin{aligned}
a &= \sum_{\sigma \in \pi} \sigma \cdot f(\sigma^{-1}a) - \sum_{\sigma \in \pi} f(\sigma^{-1}a) \\
&= \sum_{\sigma \in \pi} (\sigma - 1) \cdot f(\sigma^{-1}a) \in IA.
\end{aligned}
$$

This shows that ${}_N A \subset IA$. We know already that $IA \subset {}_N A$. Hence $IA = {}_N A$.

Let $a \in A^\pi$. Then $\sigma \cdot a = a \ \forall\ \sigma \in \pi$. From $a = \sum_{\sigma \in \pi} \sigma \cdot f(\sigma^{-1}a)$ we get $a = \sum_{\sigma \in \pi} \sigma \cdot f(a) = N(f(a))$. Thus $A^\pi \subset N(A)$. But already we know that $N(A) \subset A^\pi$. Hence $A^\pi = N(A)$.

From $IA = {}_N A$ and $A^\pi = N(A)$ we immediately see that N^* is an isomorphism and hence $\bar{N}$ and h are isomorphisms as well.

Proposition 2.14

Let $A \in Z\pi$-mod *be weakly projective. Then* $H_n(\pi, A) = 0$ *for* $n \geq 1$.

Proof. By definition, A is weakly projective $\Leftrightarrow$

$$
0 \longrightarrow \ker g \longrightarrow (Z\pi \otimes_Z A)_{(2)} \xrightarrow{\ g\ } A \longrightarrow 0 \qquad (*)
$$

is split exact in $Z\pi$-mod where $g(\lambda \otimes a) = \lambda a$ for any $\lambda \in Z\pi$ and $a \in A$. Now, $H_n(\pi, A) = \text{tor}_n^{Z\pi}(Z, A)$.

Since $(*)$ splits, if we show that $\text{tor}_n^{Z\pi}(Z, (Z\pi \otimes_Z A)_{(2)}) = 0$ for $n \geq 1$, it will follow that $\text{tor}_n^{Z\pi}(A, A) = 0$ for $n \geq 1$. Let P_* be a projective resolution of A in Z-mod. Then since $Z\pi$ is free over Z the complex $Z\pi \otimes_Z P_*$ is an "acyclic" complex over $(Z\pi \otimes_Z A)_{(2)}$ where π acts on $Z\pi \otimes_Z P_*$ through action on $Z\pi$ on the left. Also $Z\pi \otimes_Z P_*$ is a projective resolution of $(Z\pi \otimes_Z A)_{(2)}$ in $Z\pi$-mod. Hence

$$\text{tor}_n^{Z\pi}\big(A, (A\pi \otimes_Z A)_{(2)}\big) = H_n\big(Z\pi \otimes_Z P_*)\big)$$

$$= H_n\big(Z \otimes_Z P_*\big)$$

$$\simeq \text{tor}_n^Z(Z, A) = 0 \quad \text{for } n \geq 1. \qquad \square$$

Proposition 2.15

Let $A \in Z\pi$-mod be weakly injective. Then $H^n(\pi, A) = 0$ for $n \geq 1$.

Proof. By definition, A is weakly injective $\Leftrightarrow$

$$0 \longrightarrow A \xrightarrow{\ h\ } \text{Hom}_Z(Z\pi, A)_{(2)} \longrightarrow \text{coker } h \longrightarrow 0 \qquad (**)$$

is exact in $Z\pi$-mod where $\{h(a)\}(\lambda) = \lambda a \ \ \forall \ \lambda \in Z\pi$. We have $H^n(\pi, A) = \text{Ext}_{Z\pi}^n(Z, A)$. Since $(**)$ is split exact, if we show that $\text{Ext}_{Z\pi}^n(A, \text{Hom}_Z(Z\pi, A)_{(2)}) = 0$ for $n \geq 1$, it will follow that $H^n(\pi, A) = 0$ for $n \geq 1$.

For any ${}_{Z\pi}B$ and ${}_Z C$, i.e., $B \in Z\pi$-mod, $C \in Z$-mod, we have

$$\text{Hom}_{Z\pi}\big(B, \text{Hom}_Z(Z\pi, C)_{(2)}\big) \stackrel{t}{=} \text{Hom}_Z\big(A\pi \otimes_{Z\pi} B, C\big)$$

$$\simeq \text{Hom}_Z(B, C)$$

where $\{t(\varphi)\}(\lambda \otimes_{Z\pi} b) = \varphi(b)(\lambda) \ \forall \ b \in B$ and $\lambda \in Z\pi$. Suppose ${}_Z C$ is injective. Let $0 \to B' \to B$ be exact in $Z\pi$-mod. Then $0 \to B' \to B$ is exact in Z-mod as well and hence $\text{Hom}_Z(B, C) \to \text{Hom}_Z(B', C) \to 0$ is exact. It follows that $\text{Hom}_{Z\pi}(B, \text{Hom}_Z(Z\pi, C)_{(2)}) \to \text{Hom}_{Z\pi}(B', \text{Hom}_Z(Z\pi, C)_{(2)}) \to 0$ is exact. Thus $\text{Hom}_Z(Z\pi, C)_{(2)}$ is injective in $Z\pi$-mod whenever C is injective in Z-mod. Let $0 \to A \to I_0 \to I_1 \to 0 \to 0 \to \cdots$ be an exact sequence in Z-mod with I_0, I_1 injective (i.e., divisible) in Z-mod. Then since $Z\pi$ is free over Z we see that

$$0 \to \text{Hom}_Z(Z\pi, A)_{(2)} \to \text{Hom}_Z(Z\pi, I_0)_{(2)} \to \text{Hom}_Z(Z\pi, I_1)_{(2)} \to 0 \to \cdots$$

is exact in $Z\pi$-mod. Thus $\text{Hom}_Z(Z\pi, I_*)_{(2)}$ is an injective resolution of

$\operatorname{Hom}_Z(Z\pi, A)_{(2)}$ in $Z\pi$-mod. Now

$$\operatorname{Ext}^n_{Z\pi}\big(A, \operatorname{Hom}_Z(Z\pi, A)_{(2)}\big) = H^n\big(\operatorname{Hom}_{Z\pi}\big(A, \operatorname{Hom}_Z(Z\pi, I_*)_{(2)}\big)\big)$$

$$\simeq H^n\big(\operatorname{Hom}_Z(Z, I_*)\big)$$

$$\simeq \operatorname{Ext}^n_Z(Z, A) = 0 \quad \text{for } n \geq 1.$$

Thus $\operatorname{Ext}^n_{Z\pi}(Z, \operatorname{Hom}_Z(Z\pi, A)_{(2)}) = 0$ for $n \geq 1$ and hence $H^n(\pi, A) = 0$ for $n \geq 1$. □

Theorem 2.16

Let $A \in K\pi$-mod. Then the following conditions are equivalent:

(i) A is weakly projective in $K\pi$-mod.
(ii) A is weakly injective in $K\pi$-mod.
(iii) $\hat{H}(\pi, \operatorname{Hom}_K(A, A)) = 0$.

Proof. The equivalence of (i) and (ii) is already proved in Theorem 2.9.

Assume $A \in K\pi$-mod weakly projective. From Corollary 2.12 we see that $\operatorname{Hom}_K(A, A)$ is weakly projective in $K\pi$-mod and hence weakly injective as well. From Remark 2.10 we see that $\operatorname{Hom}_K(A, A)$ is weakly projective and hence weakly injective in $Z\pi$-mod. From Propositions 2.13–2.15 we get $\hat{H}(\pi, \operatorname{Hom}_K(A, A)) = 0$. This gives (i) $\Rightarrow$ (iii).

Conversely, assume that $\hat{H}(\pi, \operatorname{Hom}_K(A, A)) = 0$. In particular $\hat{H}^0(\pi, \operatorname{Hom}_K(A, A)) = 0$ or $N(\operatorname{Hom}_K(A, A)) = \operatorname{Hom}_K(A, A)^\pi$. It is clear that $\operatorname{Id}_A \in \operatorname{Hom}_K(A, A)^\pi$. In fact $\forall \, \sigma \in \pi$ and $a \in A$, we have $\{\sigma \operatorname{Id}_A\}(a) = \sigma \cdot \operatorname{Id}_A(\sigma^{-1}a) = \sigma \cdot \sigma^{-1}a = a$. Hence $\sigma \operatorname{Id}_A = \operatorname{Id}_A$. It follows that $\exists$ an $f \in \operatorname{Hom}_K(A, A)$ with $Nf = \operatorname{Id}_A$. Hence A is weakly projective. □

Remark 2.17

Let $A \in K\pi$-mod. The proof of Theorem 2.16 actually shows that

$$\hat{H}^0\big(\pi, \operatorname{Hom}_K(A, A)\big) = 0 \Leftrightarrow A \quad \text{is weakly } K\pi \text{ projective}$$

$$\Leftrightarrow \hat{H}\big(\pi, \operatorname{Hom}_K(A, A)\big) = 0.$$

Proposition 2.18

Let $A \in K\pi$-mod. Then

(i) A is projective in $K\pi$-mod $\Leftrightarrow$ A is weakly projective in $K\pi$-mod *and* A is projective in K-mod.

(ii) *A is injective in $K\pi$-mod $\Leftrightarrow$ A is weakly injective in $K\pi$-mod and A is injective in K-mod.*

Proof. (i) If A is projective in $K\pi$-mod, since $K\pi$ is free over K it follows that A is projective in K-mod. Also A in $K\pi$-mod is weakly projective.

Conversely, let A be weakly projective in $K\pi$-mod and projective in K-mod. Then A is isomorphic to a direct summand of $(K\pi \otimes_K A)_{(2)}$ that is projective in $K\pi$-mod. Hence A is projective in $K\pi$-mod.

(ii) If A is injective in $K\pi$-mod, then clearly A is weakly injective in $K\pi$-mod. Let $0 \to B' \to B$ be exact in K-mod. Since $K\pi$ is free over K, we see that $0 \to K\pi \otimes_K B' \to K\pi \otimes_K B$ is exact in $K\pi$-mod. Since A is injective in $K\pi$-mod, we see that

$$\mathrm{Hom}_{K\pi}\left(K\pi \underset{K}{\otimes} B, A\right) \to \mathrm{Hom}_{K\pi}\left(K\pi \underset{K}{\otimes} B', A\right) \to 0 \quad \text{is exact.}$$

But

$$\mathrm{Hom}_{K\pi}\left(K\pi \underset{K}{\otimes} B, A\right) \simeq \mathrm{Hom}_{K\pi}(K\pi, \mathrm{Hom}_K(B, A))$$

$$\simeq \mathrm{Hom}_K(B, A).$$

Thus $\mathrm{Hom}_K(B, A) \to \mathrm{Hom}_K(B', A) \to 0$ is exact. Hence A is injective in K-mod.

Conversely, let A be injective in K-mod and weakly injective in $K\pi$-mod. Then A is a direct summand of $\mathrm{Hom}_K(K\pi, A)_{(2)}$ and $\mathrm{Hom}_K(K\pi, A)_{(2)}$ is injective as a $K\pi$ module. Hence A is injective in $K\pi$-mod. $\square$

Corollary 2.19

Let K be a field and $A \in K\pi$-mod. The following are equivalent:

(i) *A is weakly projective in $K\pi$-mod.*

(ii) *A is weakly injective in $K\pi$-mod.*

(iii) *A is projective in $K\pi$-mod.*

(iv) *A is injective in $K\pi$-mod.*

3   *p* GROUPS

Let p be a prime and $Z_p = Z/pZ$.

Lemma 3.1

Let π be any group, π' a normal subgroup of π, and $\eta\colon \pi \to \pi/\pi'$ the canonical quotient map. Let K be any commutative ring, $\eta_*\colon K(\pi) \to K(\pi/\pi')$ the map induced by η, and $I(\pi')$, $I(\pi)$, $I(\pi/\pi')$ the augmentation ideals in $K(\pi')$, $K(\pi)$, $K(\pi/\pi')$, respectively. Then $K(\pi) \cdot I(\pi') = I(\pi') \cdot K(\pi)$ and

$$0 \longrightarrow K(\pi) \cdot I(\pi') \longrightarrow I(\pi) \stackrel{\eta_*}{\longrightarrow} I(\pi/\pi') \longrightarrow 0$$

is exact.

Proof. Since π' is a normal subgroup of π, for any $\sigma \in \pi$ and $x' \in \pi'$, we have $\sigma x' = y' \sigma$ for some $y' \in \pi'$. Thus $\sigma(x' - 1) = (y' - 1)\sigma$. This yields $K(\pi)I(\pi') = I(\pi')K(\pi)$. Let us now determine the kernel of $\eta_*\colon K(\pi) \to K(\pi/\pi')$. Let $\sigma_1, \ldots, \sigma_r$ be representatives for the distinct cosets of π' in π. Then any element a of $K(\pi)$ could be written as

$$\sum_{x' \in \pi'} \alpha_{\sigma_1 x'} \sigma_1 x' + \cdots + \sum_{x' \in \pi'} \alpha_{\sigma_r x'} \sigma_r x' \text{ with } \alpha_{\sigma_j x'} \in K, \quad 1 \le j \le r,$$

$$\eta_*(a) = 0 \text{ in } K(\pi/\pi') \Leftrightarrow \sum_{x' \in \pi'} \alpha_{\sigma_j x'} = 0 \quad \text{for } 1 \le j \le r$$

$$\Leftrightarrow \sum_{x' \in \pi'} \alpha_{\sigma_j x'} x' \in I(\pi') \quad \text{for } 1 \le j \le r.$$

Hence

$$\eta_*(a) = 0 \Rightarrow a = \sigma_1 c_1 + \cdots + \sigma_r c_r \quad \text{with } c_j \in I(\pi')$$

$$\Rightarrow a \in K(\pi) \cdot I(\pi').$$

Conversely, it is clear that $\eta_*(K(\pi) \cdot I(\pi')) = 0$. Thus $K(\pi) \cdot I(\pi') = I(\pi') \cdot K(\pi) = \ker \eta_*\colon K(\pi) \twoheadrightarrow K(\pi/\pi')$. From the commutative diagram

$$
\begin{array}{ccccccccc}
0 & \longrightarrow & K(\pi) \cdot I(\pi') & \longrightarrow & K(\pi) & \stackrel{\eta_*}{\longrightarrow} & K(\pi/\pi') & \longrightarrow & 0 \\
 & & \downarrow & & \downarrow{\scriptstyle \varepsilon} & & \downarrow{\scriptstyle \varepsilon} & & \\
0 & \longrightarrow & 0 & \longrightarrow & K & \Longrightarrow & K & \longrightarrow & 0
\end{array}
$$

we see that the sequence of kernels (of the vertical maps)

$$0 \to K(\pi) \cdot I(\pi') \to I(\pi) \to I(\pi/\pi') \to 0$$

is exact. $\square$

Lemma 3.2

Let π be a p group. Then the augmentation ideal $I(\pi)$ in $Z_p(\pi)$ is nilpotent. In particular $Z_p(\pi)$ is a local ring with $J(Z_p(\pi)) = I(\pi)$ (J is the Jacobson radical).

Proof. Let $O(\pi) = p^k$. We will prove the lemma by induction on k. If $k = 0$, $\pi = \{1\}$, $I\pi = 0$, and $Z_p(\pi) = Z_p$ is a field, there is nothing to prove. Let $k = 1$ and x be a generator for the cyclic group π of order p. Since $Z_p(\pi)$ is of characteristic p we have

$$(x - 1)^p = x^p - 1 \quad \text{in } Z_p(\pi)$$

$$= 0 \quad \text{since } x^p = 1.$$

$I(\pi)$ is generated by $(x - 1)$ over $Z_p(\pi)$ and $Z_p(\pi)$ is commutative. Hence $I(\pi)^p = 0$. Thus the lemma is valid for $k = 1$.

Let $k > 1$ and assume the lemma is valid whenever $O(\pi) = p^\ell$ with $\ell < k$. Since $O(\pi) = p^k$ with $k > 1$, the center π'' of π is nontrivial. We can definitely pick a subgroup π' of the center of π with $O(\pi') = p^d$ with $1 \le d < k$. Then $O(\pi/\pi') = p^{k-d}$ and $1 \le k - d < k$. Hence by the inductive assumption, $I(\pi/\pi')$ and $I(\pi')$ are nilpotent where $I(\pi') = \ker \varepsilon: Z_p(\pi') \to Z_p$ and $I(\pi/\pi') = \ker \varepsilon: Z_p(\pi/\pi') \to Z_p$.

Lemma 3.1 yields the exact sequence

$$0 \longrightarrow Z_p(\pi) \cdot I(\pi') \longrightarrow I(\pi) \overset{\eta_*}{\longrightarrow} I(\pi/\pi') \longrightarrow 0. \qquad (*)$$

From $Z_p(\pi) \cdot I(\pi') = I(\pi')Z_p(\pi)$ and the nilpotency of $I(\pi')$ we see that $Z_p(\pi)I(\pi')$ is nilpotent. Now $(*) \Rightarrow I(\pi)$ is nilpotent. $\square$

Proposition 3.3

Let π be a p group. Then the following are equivalent for $A \in Z_p(\pi)$-mod:

(i) *A is weakly projective in $Z(\pi)$-mod.*
(ii) *A is weakly projective in $Z_p(\pi)$-mod.*
(iii) *A is projective in $Z_p(\pi)$-mod.*
(iv) *A is injective in $Z_p(\pi)$-mod.*
(v) *A is free in $Z_p(\pi)$-mod.*
(vi) *$\hat{H}^i(\pi, A) = 0$ for some integer i.*

Proof. Since $\operatorname{Hom}_Z(A, A) = \operatorname{Hom}_{Z_p}(A, A)$, the equivalence (i) $\Leftrightarrow$ (ii) is immediate from Theorem 2.16. The equivalences (ii) $\Leftrightarrow$ (iii) $\Leftrightarrow$ (iv) follow from Corollary 2.19.

(iii) $\Leftrightarrow$ (v) is an immediate consequence of the fact that $Z_p(\pi)$ is a local ring (Lemma 3.2).

We have already seen that (i) $\Rightarrow$ (vi) (Propositions 2.13–2.15). We will prove that (vi) $\Rightarrow$ (iii). This will prove Proposition 3.3. First assume that $\hat{H}^{-2}(\pi, A) = 0$.

Since $Z_p(\pi)$ is a local ring, A admits of a projective cover $P \xrightarrow{h} A$. Let $0 \to K \to P \xrightarrow{h} A \to 0$ (i.e., $K = \ker h$). Then $K \subset J(P) = J \cdot P = I(\pi) \cdot P$. Hence

$$N(K) \subset N(I(\pi) \cdot P) = 0. \text{ Thus } {}_N K = K. \tag{1}$$

From the exact sequence

$$\hat{H}^{-2}(\pi, A) \longrightarrow \hat{H}^{-1}(\pi, K) \longrightarrow \hat{H}^{-1}(\pi, P)$$
$$\Big\| \qquad\qquad\qquad\qquad\qquad\qquad \Big\|$$
$$0 \qquad\qquad\qquad\qquad\qquad\qquad\quad 0$$

we get $\hat{H}^{-1}(\pi, K) = 0$. Hence ${}_N K = IK$ or $K = IK$ from (1). But I is nilpotent, say $I^\ell = 0$. Hence $K = KL = \cdots = I^\ell K = 0$ or $h: P \simeq A$. Thus A is projective in $Z_p(\pi)$-mod.

Now, suppose $\hat{H}^i(\pi, A) = 0$ for some $i \in Z$. We will prove that $\exists$ a $Z_p(\pi)$ module B with $\hat{H}^n(\pi, A) \simeq \hat{H}^{n-i-2}(\pi, B)$ (Lemma 3.4) for all $n \in Z$. Then $\hat{H}^{-2}(\pi, B) = \hat{H}^i(\pi, A) = 0$. Hence from what we have proved B is projective in $Z_p(\pi)$-mod. Hence $\hat{H}(\pi, B) = 0$. In particular $\hat{H}^{-2}(\pi, A) = \hat{H}^{-i-4}(\pi, B) = 0$. Hence by what has been proved already, A is projective in $Z_p(\pi)$-mod. $\square$

Lemma 3.4

Let π be a finite group and $A \in Z\pi$-mod. Then $\exists$ B, C in $Z\pi$-mod with $\hat{H}^{k+1}(\pi; C) \simeq \hat{H}^k(\pi; A) \simeq \hat{H}^{k-1}(\pi; B)$ for all $k \in Z$.

Proof. Let $0 \to C \to P \to A \to 0$ and $0 \to A \to E \to B \to 0$ be exact with P projective in $Z\pi$-mod and $\exists$ injective in $Z\pi$-mod. Then from Theorem 2.16 we see that P and E are both weakly projective and weakly injective in $Z\pi$-mod. Hence $\hat{H}^i(\pi, P) = 0 = \hat{H}^i(\pi, E)$ for all i.

The exact sequences (Theorem 1.10) associated with the two sequences

$$0 \to C \to P \to A \to 0,$$

$$0 \to A \to E \to B \to 0,$$

now yield

$$\hat{H}^k(\pi, A) \simeq \hat{H}^{k+1}(\pi, C),$$
$$\hat{H}^k(\pi, B) \simeq \hat{H}^{k+1}(\pi, A),$$

for all $k \in Z.$ □

Proposition 3.5

Let π be a p group, $A \in Z\pi$-mod. Suppose $\hat{H}^i(\pi, A) = 0 = \hat{H}^{i+1}(\pi, A)$ for some particular $i \in Z$. If A is Z-free, then A is projective in $Z\pi$-mod. If A is divisible in Z-mod, then A is injective in $Z\pi$-mod.

Proof. Let $A_p = A/_pA$ and $_pA = \ker: A \overset{p}{\to} A.$

(i) Assume A free in Z-mod. Then $0 \to A \overset{p}{\to} A \to A_p \to 0$ is exact in $Z\pi$-mod. The exactness of $\cdots \to \hat{H}^i(\pi, A) \to \hat{H}^i(\pi, A_p) \to \hat{H}^{i+1}(\pi, A) \to \cdots$ yields $\hat{H}^i(\pi, A_p) = 0$. Clearly A_p is a $Z_p\pi$-module.

From Proposition 3.3 we see that A_p is weakly projective in $Z\pi$-mod. Since A is free in Z-mod,

$$0 \longrightarrow \mathrm{Hom}_Z(A, A) \overset{p}{\longrightarrow} \mathrm{Hom}_Z(A, A) \longrightarrow \mathrm{Hom}_Z(A, A_p) \longrightarrow 0 \quad (**)$$

is exact (clearly in $Z\pi$-mod). From Proposition 2.11 we see that $\mathrm{Hom}_Z(A, A_p)$ is weakly projective in $Z\pi$-mod. Hence $\hat{H}^i(\pi, \mathrm{Hom}_Z(A, A_p)) = 0$ for all i. The exactness of $(**)$ now yields

$$\hat{H}^i(\pi, \mathrm{Hom}_Z(A, A)) \overset{p_*}{\longrightarrow} \hat{H}^i(\pi, \mathrm{Hom}_Z(A, A))$$

an isomorphism. Let $n = O(\pi)$. We have $n = p^k$ for some $k \geq 0$. Hence $\hat{H}^i(\pi, \mathrm{Hom}_Z(A, A)) \overset{n_*}{\underset{\simeq}{\longrightarrow}} \hat{H}^i(\pi, \mathrm{Hom}_Z(A, A))$. We will prove that $n_* = 0$ (Lemma 3.10). Hence $\hat{H}^i(\pi, \mathrm{Hom}_Z(A, A)) = 0$ for all i. From Theorem 2.16 it follows that A is weakly projective in $Z\pi$-mod. From Proposition 2.18, since A is free in Z-mod we see that A is projective in $Z\pi$-mod.

The dual, when A is divisible is proved similarly using the exact sequences

$$0 \to {}_pA \to A \overset{p}{\to} A \to 0,$$

$$0 \to \mathrm{Hom}_Z(A, A) \to \mathrm{Hom}_Z(A, A) \to \mathrm{Hom}_Z({}_pA, A) \to 0.$$ □

Lemma 3.6

Let $M \in K$-mod and $A \in K\pi$-mod with A weakly projective. Then $A \otimes_K M$ is weakly projective in $K\pi$-mod with π operation given by $\sigma \cdot (a \otimes x) = \sigma a \otimes x$ $\forall$ $\sigma \in \pi$, $a \in A$, and $x \in M$.

Proof. If $g: (K\pi \otimes_K A)_{(2)} \xrightarrow{g} A$ is the map $g(\lambda \otimes a) = \lambda a$, then $0 \to \ker g$ $\to (K\pi \otimes_K A)_{(2)} \xrightarrow{g} A \to 0$ is split exact in $K\pi$-mod. Let $A \xrightarrow{s} (K\pi \otimes_K A)_{(2)}$ denote a splitting of g. Then $s \otimes 1_M$ splits the map

$$\left(K\pi \otimes_K (A \otimes_K M)\right)_{(2)} \xrightarrow{g_{A \otimes_K M}} A \otimes_K M$$

where $g_{A \otimes_K M}(\lambda \otimes u) = \lambda u \ \forall \ u \in A \otimes_K M$. $\square$

Corollary 3.7

(i) *For any $M \in K$-mod, $K\pi \otimes_K M$ is weakly projective in $K\pi$-mod.*

(ii) *For any $B \in K\pi$-mod, $(K\pi \otimes_K B)_{(2)}$ is weakly projective.*

Lemma 3.8

Let B, C be in $K\pi$-mod and $f \in \mathrm{Hom}_K(B, C)$. Then $f = N(\varphi)$ for some $\varphi \in \mathrm{Hom}_K(B, C) \Leftrightarrow f$ can be factored as $v \circ u$ with

$$B \xrightarrow{u} (k\pi \otimes_K C)_{(2)}, \quad (K\pi \otimes_K C)_{(2)} \xrightarrow{v} C \quad \text{maps in } K\pi\text{-mod.}$$

Proof. Let $f = v \circ u$. Since $(K\pi \otimes_K C)_{(2)}$ is weakly projective in $K\pi$-mod (Corollary 3.7), we have a $\rho \in \mathrm{End}_K((K\pi \otimes_K C)_{(2)})$ with $N\rho = \mathrm{Id}_{(K\pi \otimes_K C)_{(2)}}$. Also

$$N(v \circ \rho \circ u) = v \circ N\rho \circ u \quad \text{since } u, v \text{ are in } K\pi\text{-mod}$$

$$= v \circ u = f.$$

Conversely, suppose $f = N(\varphi)$ with $\varphi \in \mathrm{Hom}_K(B, C)$. Let

$$u(b) = \sum_{\sigma \in \pi} \sigma \otimes \varphi(\sigma^{-1} b),$$

$$v(\lambda \otimes c) = \lambda c \quad \forall \ \lambda \in K\pi.$$

We already know that v is a $K\pi$ homomorphism.
Also for any $\tau \in \pi$,

$$u(\tau b) = \sum_{\sigma \in \pi} \sigma \otimes \varphi(\sigma^{-1} \tau b)$$

$$= \sum_{\sigma \in \pi} \tau \sigma' \otimes \varphi(\sigma'^{-1} b) \quad (\sigma' = \tau^{-1}\sigma)$$

$$= \tau u(b).$$

Thus $u \in \text{Hom}_{K\pi}(B, (K\pi \otimes_K C)_{(2)})$. Moreover $v \circ u(b) = \Sigma_{\sigma \in \pi}\sigma \cdot \varphi(\sigma^{-1}b)$ $= (N\varphi)(b) = f(b)$. Thus $f = v \circ u$. Observe that $N(\text{Hom}_K(B, C)) \subset$ $\text{Hom}_{K\pi}(B, C)$. $\square$

Lemma 3.9

Let $B \xrightarrow{f} C$ be a map in $K\pi$-mod. Suppose $f \in N(\text{Hom}_K(B, C))$. Then $f_*: \hat{H}(\pi, B) \to \hat{H}(\pi, C)$ is the zero homomorphism.

Proof. By Lemma 3.8, $f = v \circ u$ with $u: B \to (K\pi \otimes_K C)_{(2)}$ and $v: (K\pi \otimes_K C)_{(2)} \to C$ maps in $K\pi$-mod. Since $(K\pi \otimes_K C)_{(2)}$ is weakly projective (Corollary 3.7) we get $\hat{h}(\pi, (K\pi \otimes_K C)_{(2)}) = 0$. From $f_* = v_* \circ u_*$ we now get $f_* = 0$. $\square$

Lemma 3.10

Let $O(\pi) = n$ and $A \in Z\pi$-mod. Then $n_*: \hat{H}(\pi, A) \to \hat{H}(\pi, A)$ is the zero map.

Proof. The map $A \xrightarrow{n} A$ is $N(\text{Id}_A)$. Hence Lemma 3.9 yields $n_* = 0$. $\square$

4 COHOMOLOGICALLY TRIVIAL MODULES

Definition 4.1

A π-module A will be called *cohomologically trivial* if $\hat{H}(\pi', A) = 0$ for all subgroups π' of π.

If $O(\pi) = n$ we have already seen that $n_*: \hat{H}(\pi, A) \to \hat{H}(\pi, A)$ is the zero map (Lemma 3.10). Hence $\hat{H}(\pi, A)$ is a torsion abelian group with the property that the order of any nonzero element of $\hat{H}(\pi, A)$ is a divisor of n. For any prime p, let $\hat{H}(\pi, A, p)$ denote the p-primary component of $\hat{H}(\pi, A)$. Then $\hat{H}(\pi, A, p) = 0$ for $p \times n$ and $\hat{H}(\pi, A) = \oplus_{p|n} \hat{H}(\pi, A, p)$. Also when $p|n$, write $n = p^k d$ with $(p, d) = 1$. Then every nonzero element of $\hat{H}(\pi, A, p)$ has an order dividing p^k or $p^k \hat{H}(\pi, A, p) = 0$.

Given a subgroup G of π, there exist maps

$$i(G, \pi): \hat{H}(\pi, A) \to \hat{H}(G, A),$$

$$t(\pi, G): \hat{H}(G, A) \to \hat{H}(\pi, A),$$

called the restriction and transfer maps (see [4], Chapter 12). The only property of these maps that we need is

$$\forall\, x \in \hat{H}(\pi, A), \qquad t(\pi, G)i(G, \pi)(x) = (\pi : G)x. \qquad (4.2)$$

Lemma 4.3

Let $A \in Z\pi$-mod and π' be a Sylow p subgroup of π. Then

$$t(\pi, \pi'): \hat{H}(\pi', A) \to \hat{H}(\pi, A, p)$$

is an epimorphism.

Proof. Since $\hat{H}(\pi', A) = \hat{H}(\pi', A, p)$ it is clear that $t(\pi, \pi')(\hat{H}(\pi', A)) \subset \hat{H}(\pi, A, p)$. Let $O(\pi) = n$ and $O(\pi') = p^k$. Then $n = p^k d$ with $(p, d) = 1$ and $(\pi : \pi') = d$. Since $(p^k, d) = 1$ we can find an integer ℓ such that $\ell d \equiv 1$ (mod p^k). Let $x \in \hat{H}(\pi, A, p)$. Then $p^k x = 0$.

Also

$$\ell t(\pi, \pi')i(\pi', \pi)(x) = \ell(\pi : \pi')x$$

$$= \ell\, dx.$$

But $\ell d = 1 + rp^k$ with $r \in Z$. Thus $\ell\, dx = x + rp^k x = x$. This shows that $t(\pi, \pi')(\ell i(\pi', \pi)(x)) = x$. Hence $t(\pi, \pi'): \hat{H}(\pi', A) \to \hat{H}(\pi, A, p)$ is onto. $\qquad\square$

Corollary 4.4

If $\hat{H}(\pi; A) = 0$ for all Sylow subgroups π' of π, then $\hat{H}(\pi, A) = 0$.

Proposition 4.5

Let $A \in Z\pi$-mod. Then the following are equivalent:

  (i) *A is weakly projective in $Z\pi$-mod.*
 (ii) *A is weakly projective in $Z\pi'$-mod for any $\pi' \subset \pi$.*
(iii) *A is weakly projective in $Z\pi'$-mod for every Sylow subgroup π' of π.*

Proof. (i) $\Rightarrow$ (ii). Since A is weakly projective, $\exists$ an $f \in \mathrm{Hom}_Z(A, A)$ with $\sum_{\sigma \in \pi}\sigma \cdot f(\sigma^{-1}a) = a\ \forall\ a \in A$. Let $\tau_1, \ldots, \tau_k$ be representatives for the distinct right cosets of π' in π, namely $\pi'\tau_1, \ldots, \pi'\tau_k$ are all the distinct cosets of

π' in π. Let $\varphi \in \text{Hom}_Z(A, A)$ be given by $\varphi(a) = \sum_{i=1}^k \tau_i f(\tau_i^{-1}a)$. Then

$$\sum_{\alpha \in \pi'} \alpha \cdot \varphi(\alpha^{-1}a) = \sum_{\alpha \in \pi'} \alpha \cdot \left(\sum_{i=1}^k \tau_i f(\tau_i^{-1}\alpha^{-1}a) \right)$$

$$= \sum_{\alpha \in \pi'} \sum_{i=1}^k \alpha\tau_i f(\alpha\tau_i)^{-1}a)$$

$$= \sum_{\sigma \in \pi} \sigma \cdot f(\sigma^{-1}a) = a.$$

This proves that A is weakly $Z\pi'$ projective.

(ii) $\Rightarrow$ (iii). Trivial.

(iii) $\Rightarrow$ (i). From (iii) we get $\hat{H}(\pi', \text{Hom}_Z(A, A)) = 0$ for any Sylow subgroup π' of π (see Theorem 2.16). From Corollary 4.4 we see that $\hat{H}(\pi, \text{Hom}_Z(A, A)) = 0$. Again from Theorem 2.16 we see that A is weakly projective in $Z\pi$-mod. $\square$

Proposition 4.6

Let $A \in Z\pi$-mod. The following are equivalent:

(i) A is projective in $Z\pi$-mod.
(ii) A is projective in $Z\pi'$-mod for every $\pi' \subset \pi$.
(iii) A is projective in $Z\pi'$-mod for every Sylow subgroup π' of π.

Proof. (i) $\Rightarrow$ (ii) $\Rightarrow$ (iii) are trivial. Also if (iii) is valid, then A is Z-free and A is weakly projective in $Z\pi'$-mod for every Sylow subgroup π' of π. By Proposition 4.5, A is then weakly projective in $Z\pi$-mod. Also A is Z-free. From Proposition 2.18 we see that A is projective in $Z\pi$-mod. $\square$

The following is proved similarly.

Proposition 4.7

Let $A \in Z\pi$-mod. Then the following are equivalent:

(i) A is injective in $Z\pi$-mod.
(ii) A is injective in $Z\pi'$-mod for all $\pi' \subset \pi$.
(iii) A is injective in $Z\pi'$-mod for all Sylow subgroups π' of π.

Theorem 4.8

Let $A \in Z\pi$-mod. Then A is projective (resp. injective) in $Z\pi$-mod $\Leftrightarrow A$ is Z-free (resp. Z-divisible) and cohomologically trivial.

Proof. If A is projective (resp. injective) in $Z\pi$-mod, then A is Z-free (resp. Z-divisible) and weakly projective (equivalently weakly injective) in $Z\pi$-mod (Proposition 2.18).

From Proposition 4.5, A is weakly projective in $Z\pi'$-mod for every $\pi' \subset \pi$ and hence by Propositions 2.13–2.15 A is cohomologically trivial.

Conversely, suppose A is cohomologically trivial and Z-free (resp. Z-divisible). If π' is any Sylow subgroup of π, from Proposition 3.5 we see that A is projective (resp. injective) in $Z\pi'$-mod. Hence from Proposition 4.6 (resp. Proposition 4.7) A is projective (injective) in $Z\pi$-mod. $\square$

Theorem 4.9

Let $A \in Z\pi$-mod. Then the following are equivalent:

(i) A is cohomologically trivial.
(ii) For any prime p and any Sylow subgroup π' of π, $\exists$ an integer i_p such that $\hat{H}^{i_p}(\pi', A) = 0 = \hat{H}^{i_p+1}(\pi', A)$.
(iii) proj dim$_{Z(\pi)} A \leq 1$.
(iii′) inj dim$_{Z(\pi)} A \leq 1$.
(iv) proj dim$_{Z(\pi)} A < \infty$.
(iv′) inj dim$_{Z(\pi)} A < \infty$.

Proof.
(i) $\Rightarrow$ (ii). Trivial.
(ii) $\Rightarrow$ (iii). Let $0 \to M \to P \to A \to 0$ be exact with P projective in $Z\pi$-mod. Let π' be a Sylow p subgroup of π. Then from (ii) and the extended cohomology exact sequence of $0 \to M \to P \to A \to 0$ (as π' modules) we get $\hat{H}^{i_p+1}(\pi', M) = 0 = \hat{H}^{i_p+2}(\pi', M)$ [observe that P is projective in $Z\pi'$-mod and hence weakly projective in $Z\pi'$-mod and hence $\hat{H}(\pi', P) = 0$]. Also P is free in Z-mod and hence M is free in Z-mod. From Proposition 3.5 we see that M is projective in $Z\pi'$-mod. This is true for every Sylow subgroup π' of π. From Proposition 4.6 we see that M is projective in $Z\pi$-mod. Hence proj dim$_{Z(\pi)} A \leq 1$.
(iii) $\Rightarrow$ (iv). Trivial.
(iv) $\Rightarrow$ (i). Let proj dim$_{Z(\pi)} A = n < \infty$. Let

$$0 \to M \to P_{n-1} \to P_{n-2} \to \cdots \to P_0 \to A \to 0 \qquad (*)$$

be an exact sequence in $Z\pi$-mod with P_i projective $0 \leq i \leq n - 1$. Since proj dim$_Z(\pi)A \leq n$ we get M to be projective in $Z\pi$-mod. Since P_i are projective in $Z\pi$-mod, and hence cohomologically trivial we see that $\hat{H}^{i+n}(\pi', M) \simeq \hat{H}^i(\pi', A)$ for any $\pi' \subset \pi$. Since M itself is projective in $Z\pi$-mod, $\hat{H}^{i+n}(\pi', M) = 0$ for all i and hence $\hat{H}^i(\pi', A) = 0$ for all $i \in Z$. Thus A is cohomologically trivial.

Thus we have proved (i) $\Rightarrow$ (ii) $\Rightarrow$ (iii) $\Rightarrow$ (iv) $\Rightarrow$ (i). Similarly it can be shown that (i) $\Rightarrow$ (ii) $\Rightarrow$ (iii') $\Rightarrow$ (iv') $\Rightarrow$ (i). $\square$

5 $\tilde{K}_0(Z\pi)$ WHEN π IS A CYCLIC GROUP OF ORDER p, WITH p A PRIME

In this section π denotes a cyclic group of order p where p is a prime and x denotes a fixed generator of the group π. Let $N = \sum_{\sigma \in \pi} \sigma = \sum_{i=0}^{p-1} x^i$. From $\sigma \cdot N = N = N \cdot \sigma$ for every $\sigma \in \pi$ we see that $(N) = \{mN | m \in Z\}$ is a two-sided ideal in $Z\pi$. Thus $Z\pi \cdot N = N \cdot Z\pi = (N)$. Let $\omega = \exp(2\pi\sqrt{-1}/p) \in \mathbb{C}$. Then $\omega, \omega^2, \ldots, \omega^{p-1}$ are all the primitive pth roots of unity in $\mathbb{C}$. Let $Q(\omega)$ denote the cyclotomic extension field of Q gotten by adjoining ω to Q. $Q(\omega)$ is an algebraic number field. Since ω is a root of $X^{p-1} + X^{p-2} + \cdots + 1$, it is clear that ω is an algebraic integer. We will assume the following result.

Theorem 5.1

The ring of algebraic integers in $Q(\omega)$ is precisely $Z[\omega]$. In particular $Z[\omega]$ is a Dedekind domain.

Lemma 5.2

Let $f: Z\pi \to Z[\omega]$ be the ring homomorphism carrying $x \in \pi$ onto ω. Then ker $f = (N)$.

Proof. Let $Z[X]$ denote the polynomial ring. Any element in $Z\pi$ could be written as $\theta(x)$ with $\theta(X) \in Z[X]$. Then $f(\theta(x)) = \theta(\omega)$. Thus $f(\theta(x)) = 0$ $\Leftrightarrow \theta(\omega) = 0$. But the cyclotomic polynomial $\Phi_p(X) = X^{p-1} + X^{p-2} + \cdots + 1$ is the irreducible polynomial of ω over Q. Thus $\theta(X) = \Phi_p(X)h(X)$ with $h(X) \in Q[X]$. Now comparing coefficients and using the fact that $\theta(X) \in Z[X]$ we see that $h(X) \in Z[X]$. It now follows that $\theta(x) = \Phi_p(x)h(x) = Nh(x) \in Z\pi$. Thus $\theta(x) \in$ ker $f \Rightarrow \theta(x) \in (N)$. It is clear that $(N) \subset$ ker f.

Thus f induces an isomorphism

$$\bar{f}: \frac{Z\pi}{(N)} \simeq A[\omega].$$ $\square$

Lemma 5.3

$\prod_{\mu=1}^{p-1}(1 - \omega^{\mu}) = p$ in $Z[\omega]$.

Proof. $\omega, \omega^2, \ldots, \omega^{p-1}$ are the distinct roots of $\Phi_p(X)$ in $\mathbb{C}$. Then in $\mathbb{C}[X]$ we have $\Phi_p(X) = \prod_{\mu=1}^{p-1}(X - \omega^{\mu})$. Substitute $X = 1$ to get $p = \prod_{\mu=1}^{p-1}(1 - \omega^{\mu})$ in $\mathbb{C}$. But p and $\prod_{\mu=1}^{p-1}(1 - \omega^{\mu})$ are in $Z[\omega] \subset \mathbb{C}$. $\square$

Lemma 5.4

For any $1 \leq \mu \leq p - 1$, *the map* $Z[\omega] \to Z[\omega^{\mu}] = Z[\omega]$ *carrying* ω *to* ω^{μ} *and leaving* Z *fixed is a ring isomorphism.*

Clear.

Lemma 5.5

$1 - \omega$ *is not a unit in* $Z[\omega]$.

Proof. On the contrary, assume that $1 - \omega$ is a unit in $Z[\omega]$. Then $1 - \omega^{\mu}$ is a unit in $Z[\omega^{\mu}] = Z[\omega]$ from Lemma 5.4. Hence $p = \prod_{\mu=1}^{p-1}(1 - \omega^{\mu})$ will then be a unit in $Z[\omega]$. We have $Z[\omega] \subset Q(\omega)$ and in the field $Q(\omega)$, $1/p$ is the inverse of p. Suppose $\alpha \in Z[\omega]$ is the inverse of p in $Z[\omega]$. Then $\alpha = 1/p$ [since α will then be inverse of p in $Q(\omega)$ as well]. We know that $Z[\omega] \cap Q = Z$ (because Z is integrally closed in Q). From $\alpha = 1/p \in Q \cap Z[\omega] = Z$ we should have $1/p$ an integer, an impossibility. $\square$

Corollary 5.6

$(1 - \omega)Z[\omega] \neq Z[\omega]$.

Corollary 5.7

$(Z[\omega])/(1 - \omega) \simeq Z_p$.

In fact if $\eta: Z[\omega] \to Z[\omega]/(1 - \omega)$ denotes the canonical quotient map, we have from $0 = 1 + \omega + \omega^2 + \cdots + \omega^{p-1}$, $0 = \eta(1 + \omega + \cdots + \omega^{p-1}) = \eta(1) + \eta(1) + \cdots + \eta(1) = p\eta(1)$.

Thus $Z[\omega]/(1 - \omega)$ is a ring of characteristic p. (We know that $(1 - \omega)Z[\omega] \neq Z[\omega]$.)

Moreover any element of $Z[\omega]$ being of the form $\theta(\omega)$ with $\theta(X) \in Z[X]$, we see that $\eta(\theta(\omega)) = \eta(\theta(1)) \in \eta(Z)$. If $\gamma: Z \to Z_p$ denotes the quotient map, then $\eta/Z = \gamma$. Thus $(Z[\omega])/(1 - \omega) = \gamma(Z) \simeq Z_p$.

Denote the ring $Z[\omega]$ by R and the field $Q(\omega)$ by K.

Lemma 5.8

For any nonzero ideal A in $Z[\omega]$ we have

$$\frac{A}{(1 - \omega)A} \simeq Z_p \text{ in Z-mod.}$$

Proof. Let A^{-1} denote the inverse of A. It is a fractional ideal of R. From $AA^{-1} = R$ we get

$$(1 - \omega)AA^{-1} = (1 - \omega)R.$$

If $(1 - \omega)A = A$ we would get $(1 - \omega)AA^{-1} = AA^{-1} = R$. Thus $(1 - \omega)R = R$, contradicting Corollary 5.6. Hence $(1 - \omega)A \neq A$.

If C is any fractional ideal of R, then C is isomorphic to an integral ideal of R (in R-mod) and hence $(1 - \omega)C \neq C$.

$A/[(1 - \omega)A]$ is an $(Z[\omega])/((1 - \omega)Z[\omega])$-module. Hence $A/[(1 - \omega)A]$ is a vector space over Z_p. If we show that $A/[(1 - \omega)A]$ has dimension 1 over Z_p, we would be proving Lemma 5.8.

We know $A \oplus A^{-1} \simeq R \oplus R$ in R-mod (since R is a Dedekind domain). Hence

$$\frac{A}{(1 - \omega)A} \oplus \frac{A^{-1}}{(1 - \omega)A^{-1}} \simeq \frac{R}{(1 - \omega)R} \oplus \frac{R}{(1 - \omega)R}$$

$$\simeq Z_p \oplus Z_p.$$

We know $A/[(1 - \omega)A] \neq 0 \neq A^{-1}/[(1 - \omega)A^{-1}]$.

It now follows that

$$\dim_{Z_p} \frac{A}{(1 - \omega)A} = 1 = \dim_{Z_p} \frac{A^{-1}}{(1 - \omega)A^{-1}}. \qquad \square$$

Proposition 5.9

Let $A \neq 0$ be any ideal in $Z[\omega] = R$. Then

(i) $A/[(1 - \omega)A] \simeq Z_p$ in Z-mod.
(ii) A is free abelian of rank $p - 1$.
(iii) A as an R-module is torsion-free of rank 1 (i.e., $\dim_K K \otimes_R A = \dim_K KA = 1$). Here $K = Q(\omega)$.

Proof. (i) is proved in Lemma 5.8.

(iii) and (ii). We know that $Z[\omega]$ is noetherian and A is f.g. over $Z[\omega]$ and hence f.g. over Z. It is torsion-free over Z and hence free over Z. Also $KA = K$. Hence $\dim_K KA = 1$. This shows that the rank of A as a $Z[\omega]$-module is 1. But rank $Z[\omega]$ as a Z-module is $p - 1$. Hence rank $A = p - 1$ as an abelian group.

Let A be a nonzero ideal in $R = Z[\omega]$. Then as seen already $A/[(\omega - 1)A] \simeq Z/pZ$. Let $d \in A$ be such that $d \notin (\omega - 1)A$. For any such d we define a $Z\pi$-module A_d as

$$A_d = A \oplus Z \quad \text{as an abelian group.}$$

The action of π is given by

$$x(a, 0) = (\omega a, 0),$$

$$x^i(a, 0) = (\omega^i a, 0) \quad \text{for } i \geq 1,$$

$$x(0, 1) = (d, 1),$$

$$x^i(0, 1) = \left((1 + \omega + \cdots + \omega^{i-1})d, 1\right) \quad \text{for } i \geq 1.$$

Observe that

$$x^p(0, 1) = \left((1 + \omega + \cdots + \omega^{p-1})d, 1\right)$$

$$= (0, 1).$$

Thus we do get an action of π on A_d. We will presently show that the module A_d is cohomologically trivial. $\square$

Proposition 5.10

$$\left.\begin{aligned}\hat{H}^{2n}(\pi, Z) &\simeq Z/pZ\\ \hat{H}^{2n+1}(\pi, Z) &= 0\end{aligned}\right\} \quad \text{for all } n \in Z \tag{5.11}$$

where Z has trivial π operations.

For any nonzero ideal A in $Z[\omega]$ with π action given by $xa = \omega a \ \forall \ a \in A$,

$$\hat{H}^{2n}(\pi, A) = 0,$$

$$\hat{H}^{2n+1}(\pi, A) \simeq Z/pZ. \tag{5.12}$$

$$\hat{H}^i(\pi, A_d) = 0 \quad \text{for all } d \in A \text{ with } d \notin (\omega - 1)A. \tag{5.13}$$

Proof. Let X_* be the standard complete resolution for π given by $X_i = Z\pi$ for all integers i,

$$X_{2j} \xrightarrow{\partial_{2j}} X_{2j-1} \quad \text{the same as } N\colon Z\pi \longrightarrow Z\pi,$$

$$X_{2j+1} \xrightarrow{\partial_{2j+1}} X_{2j} \quad \text{given by } \partial_{2j+1} = (x-1)\colon Z\pi \longrightarrow Z\pi.$$

We use X_* to compute $\hat{H}(\pi; Z)$, $\hat{H}(\pi, A)$, and $\hat{H}(\pi, Z) = H(\operatorname{Hom}_\pi(X_*, Z))$. Write $C^*(X) = \operatorname{Hom}_\pi(X_*, Z)$. Then $C^i(Z) = Z$,

$$C^{2j-1}(X) \xrightarrow{\delta^{2j}} C^{2j}(Z) \quad \text{is the map } p\colon Z \longrightarrow Z,$$

$$C^{2j}(Z) \xrightarrow{\delta^{2j+1}} C^{2j+1}(Z) \quad \text{is the map } 0\colon Z \longrightarrow Z.$$

Hence $\hat{H}^{2j-1}(\pi, Z) = 0$ and $\hat{H}^{2j}(\pi, Z) = Z/pZ$.

The element $1 \in C^{2j}(Z)$ satisfies $\delta^{2j+1}(1) = 0$ and its class $[1]$ generates $\hat{H}^{2j}(\pi, Z) = Z/pZ$. This proves (5.11).

Let $A \neq 0$ be an ideal in $Z[\omega]$. Write $C^*(A)$ for $\operatorname{Hom}_\pi(X_*, A)$. Then $C^i(A) = A$ for all i,

$$C^{2j-1}(A) \xrightarrow{\delta^{2j}} C^{2j}(A) \quad \text{is the map } A \xrightarrow{0} A \text{ (observe that } {}_N A = A),$$

$$C^{2j}(A) \xrightarrow{\delta^{2j+1}} C^{2j+1}(A) \quad \text{is the map } (\omega - 1)\colon A \longrightarrow A.$$

Hence

$$\hat{H}^{2j-1}(\pi, A) = \frac{A}{(\omega - 1)A} \simeq \frac{A}{pZ},$$

$$\hat{H}^{2j}(\pi, A) = 0$$

since $A \subset Z[\omega]$, $Z[\omega]$ is an integral domain, and hence $\ker(\omega - 1): A \to A$ is 0.

The element $d \in A = C^{2j-1}(A)$ represents a generator for $A/[(\omega - 1)A] \simeq Z/pZ$. This proves (5.12).

Let Z be the trivial π-module Z and $\mu: A \to A_d$ and $\nu: A_d \to Z$ be the maps defined by

$$\mu(a) = (a, 0),$$

$$\nu(a, m) = m.$$

It is clear that μ and ν are π homomorphisms and that

$$0 \to A \xrightarrow{\mu} A_d \xrightarrow{\nu} Z \to 0$$

is an exact sequence of π-modules.

Writing $C^*(A_d)$ for $\mathrm{Hom}_\pi(X_*, A_d)$ we have the associated exact sequence

$$0 \longrightarrow C^*(A) \xrightarrow{\mu_*} C^*(A_d) \xrightarrow{\nu_*} C^*(Z) \longrightarrow 0$$

of cochain complexes of abelian groups. We know that

$$\hat{H}^{2j}(C^*(Z)) = Z/pZ, \qquad \hat{H}^{2j+1}(C^*(Z)) = 0,$$

$$\hat{H}^{2j}(C^*(A)) = 0, \qquad \hat{H}^{2j+1}(C^*(A)) \simeq Z/pZ = A/(\omega - 1)A,$$

with $1 \in C^{2j}(Z) = Z$ getting mapped to the generator of $\hat{H}^{2j}(C^*(Z))$ and $d \in C^{2j+1}(A) = A$ getting mapped to a generator of $A/[(\omega - 1)A] \simeq A/pZ$. In the diagram

$$
\begin{array}{ccc}
\downarrow & & \downarrow \\
\cdots \longrightarrow C^{2j}(A) & \xrightarrow{(\omega-1)} & C^{2j+1}(A) \longrightarrow \\
\downarrow{\scriptstyle \mu_*} & & \downarrow{\scriptstyle \mu_*} \\
\cdots \longrightarrow C^{2j}(A_d) & \xrightarrow[(x-1)]{\delta} & C^{2j+1}(A_d) \longrightarrow \\
\downarrow{\scriptstyle \nu_*} & & \downarrow{\scriptstyle \nu_*} \\
\cdots \longrightarrow C^{2j}(Z) & \xrightarrow[0]{\delta} & C^{2j+1}(Z) \longrightarrow \\
\downarrow & & \downarrow \\
0 & & 0
\end{array}
$$

the element $(0, 1) \in C^{2j}(A_d) = A_d$ satisfies $\nu_*(0, 1) = 1 \in C^{2j}(Z) = Z$. Also $\delta^{2j+1}(0, 1) = (x - 1)(0, 1) = (d, 1) - (0, 1) = (d, 0) \in A_d = C^{2j+1}(A_d)$. The element $d \in C^{2j+1}(A) = A$ satisfies $\mu_*(d) = (d, 0) = \delta^{2j+1}(0, 1)$. From the definition of $\delta: \hat{H}^{2j}(\pi, Z) \to \hat{H}^{2j+1}(\pi, A)$ we have $\delta([1]) = [d]$. Thus $\delta: \hat{H}^{2j}(\pi, Z) = Z/pZ \to \hat{H}^{2j+1}(\pi, A) = Z/pZ$ is an isomorphism.

The exact "extended cohomology" sequence associated to

$$0 \to A \xrightarrow{\mu} A_d \xrightarrow{\nu} Z \to 0$$

now yields $\hat{H}(\pi, A_d) = 0$. □

We will need the following result owing to Reiner [16].

Proposition 5.14

Let M be any finitely generated $Z\pi$-module with M torsion-free as a Z-module. Then

$$M \simeq A_{d_1}^{(1)} \oplus \cdots \oplus A_{d_r}^{(r)} \oplus B_2 \oplus \cdots \oplus B_s \oplus \frac{Z \oplus \cdots \oplus Z}{t \text{ copies}}$$

for some nonzero ideals $A^{(i)}$, B_j in $Z[\omega]$, $d_i \in A^{(i)}$, and $d_i \notin (\omega - 1)A^{(i)}$. Complete invariants for M are the integers s, t equal to the rank of A over Z (written as $[A : Z]$) and the ideal class of $_N M$ in $\mathrm{Cl}(Z[\omega])$.

We need to explain, what we mean by the ideal class of $_N M$ in $\mathrm{Cl}(Z[\omega])$. First of all, for any ideal $A \neq 0$ in $Z[\omega]$ and any $d \in A$ and $d \notin (\omega - 1)A$ we have

$$_N A_d = A.$$

We have $_N A_d \supset IA_d$. Also $(x - 1)(a, 0) = ((\omega - 1)a, 0)$ and $(x - 1)(0, 1) = (d, 0)$. It follows that $IA_d \subset A$. Also $IA_d \supset (\omega - 1)A + Z[\omega]d$. Since $A/[(\omega - 1)A] \simeq Z/pZ$ with $[d]$ generating Z/pZ, we get

$$IA_d = (\omega - 1)A + Z[\omega]d = A.$$

But $\hat{H}(\pi, A_d) = 0 \Rightarrow {}_N A_d = IA_d$. Hence $_N A_d = A$.

It follows from the preceding comments that

$$_N M = A^{(1)} \oplus \cdots \oplus A^{(r)} \oplus B_1 \oplus \cdots \oplus B_s.$$

This ideal class of $_N M$ is $A^{(1)} \cdots A^{(r)}B_1 \cdots B_s$ in $\mathrm{Cl}(Z[\omega])$.

Proposition 5.15

Let P be a finitely generated projective $Z\pi$-module. Then the isomorphism class of P is completely determined by the ideal class of ${}_N P$ in $\mathrm{Cl}(Z[\omega])$ and $[P:Z]$.

In particular if P_1, P_2 are f.g. projective $Z\pi$-modules, then $P_1 \simeq P_2 \Leftrightarrow {}_N P_1 \simeq {}_N P_2$ in $Z[\omega]$-mod.

Proof. π is a cyclic group of order p. Hence a finitely generated Z-torsion-free π-module M is projective in $Z\pi$-mod $\Leftrightarrow \hat{H}(\pi, M) = 0$ (Theorem 4.8). Let

$$M \simeq A_{d_1}^{(1)} \oplus \cdots \oplus A_{d_r}^{(r)} \oplus B_1 \oplus \cdots \oplus B_s \oplus \frac{Z \oplus \cdots \oplus Z}{t \text{ copies}}$$

be the decomposition given by Proposition 5.14. From Proposition 5.10 we see that $\hat{H}(\pi, M) = 0 \Leftrightarrow s = t = 0$, in which case $M \simeq A_{d_1}^{(1)} \oplus \cdots \oplus A_{d_r}^{(r)}$. Proposition 5.14 now yields the first part of Proposition 5.15.

If $P \simeq A_{d_1}^{(1)} \oplus \cdots \oplus A_{d_r}^{(r)}$, then ${}_N P \simeq A^{(1)} \oplus \cdots \oplus A^{(r)}$ and ${}_N P$ is projective in $Z[\omega]$-mod (since $Z[\omega]$ is hereditary). The isomorphism class of ${}_N P$ in $Z[\omega]$-mod is completely determined by the class of $A^{(1)} \cdots A^{(r)}$ in $\mathrm{Cl}(Z[\omega])$ and the integer r (Chapter 4, Theorem 3.32). Now $[{}_N P : Z] = r(p - 1)$ by Proposition 5.9. Hence $[P : Z] = r + (p - 1)r = pr$. Thus the integer r is completely determined by $[P : Z]$ and determines $[P : Z]$. Let P, P' be f.g. in projective in $Z\pi$-mod. Then $P \simeq P'$ in $Z\pi$-mod $\Leftrightarrow {}_N P$ and ${}_N P'$ determine the same element in $\mathrm{Cl}[Z[\omega]]$ and $[P : Z] = [P' : Z]$. But $[{}_N P : Z] = [P : Z](p - 1)/p = [P' : Z](p - 1)/p = [{}_N P' : Z]$. Hence $P \simeq P'$ in $Z\pi$-mod $\Leftrightarrow {}_N P \simeq {}_N P'$ in $Z[\omega]$-mod. $\square$

Corollary 5.16

$f_*: \tilde{K}_0(Z\pi) \simeq \tilde{K}_0(Z[\omega]) \simeq \mathrm{Cl}(Z[\omega])$.

Chapter Six

Finiteness Obstruction of
C. T. C. Wall

1 HOMOTOPY DOMINATION

All the spaces considered will be arcwise-connected. The only exception will
be when we consider the 0 skeleton of a CW-complex. Recall the following
definition already introduced in Chapter 1.

Definition 1.1

We say that X is dominated by Y if there exist maps

$$f: X \to Y, \qquad g: Y \to X \quad \text{with } gf \sim \text{Id}_X.$$

From Theorem 3.9 of Chapter 1 we see any space X dominated by a
CW-complex L is of the homotopy type of some CW-complex. But when X is
dominated by a *finite* CW-complex L, the question of whether X is of the
homotopy type of a finite complex remained unsolved for a long time before
Wall settled it negatively in his very original paper [24].

We first fix certain notations. If X admits a universal covering, we denote
the universal covering of X by $\tilde{X}$. However there may be some occasions when
for a subspace A of X we will write $\tilde{A}$ for $p^{-1}(A)$ where $p: \tilde{X} \to X$ is the
covering projection. In such a situation $\tilde{A}$ will not be the universal covering of
A in general. When we deal with such a situation we will explicitly mention it.
Also for any CW-complex K and any integer $k \geq 0$ we write K^k for the k
skeleton of K. We will write $K = K^k$ to mean that dim $K \leq k$. (When
$K = K^k$, it is not necessary that dim K is actually k; it can be less than k.) We
are interested in mainly dealing with problems concerning spaces having the

homotopy type of CW-complexes satisfying certain restrictions. *Thus it will be assumed that all the spaces we consider will be of the homotopy type of a CW-complex.*

Definition 1.2

A CW-complex K is said to be of finite type if K has finitely many cells in each dimension, or equivalently K^n is a finite complex for each integer $n \geq 0$.

A general construction, originally owing to John Milnor will be repeatedly used in our study of finiteness conditions for CW-complexes. We now deal with this construction.

CONSTRUCTION OWING TO J. W. MILNOR (Unpublished, but described in [24])

Let $K \xrightarrow{\varphi} X$ be an $(n-1)$-connected map with $n \geq 2$. If $n \geq 3$, we know that $\varphi_*: \pi_1(K) \simeq \pi_1(X)$. When $n \geq 3$, let $\tilde{\varphi}: \tilde{K} \to \tilde{X}$ denote a lift of φ to the universal coverings. Then $\pi_n(\tilde{\varphi}) \simeq \pi_n(\varphi)$ is a module over $Z\pi$ where $\pi = \pi_1(X)$. In case $n = 2$, let $\{\alpha_j\}_{j \in J}$ denote a set of generators for the group $\pi_2(\varphi)$. If $n \geq 3$, let $\{\alpha_j\}_{j \in J}$ denote a set of generators for the module $\pi_n(\tilde{\varphi}) \simeq \pi_n(\varphi)$ over $Z\pi$. Let

$$
\begin{array}{ccc}
S^{n-1} & \xrightarrow{f_j} & K \\
\downarrow & & \downarrow{\varphi} \\
D^n & \xrightarrow{g_i} & X
\end{array}
$$

represent $\alpha_j \in \pi_n(\varphi)$. Let $L = K \cup_{f_j} \{e_j^n\}_{j \in J}$. Then there exists an obvious extension $\Psi: L \to X$ of φ, where $\Psi | e_j^n$ arises from g_j.

In case J is finite, L is obtained from K by attaching a finite number of n cells.

Lemma 1.3

The map $\Psi: L \to X$ is n-connected.

Proof. Let a and b be the maps described in Diagram A.

$$
\begin{array}{ccc}
K = K & \xrightarrow{i} & L \\
\downarrow{i} \;\; a & \downarrow{\varphi} \;\; b & \downarrow{\Psi} \\
L \xrightarrow[\Psi]{} & X = X &
\end{array}
$$

Diagram A.

where $i: K \to L$ denotes the inclusion. Then we have an exact sequence

$$\cdots \longrightarrow \pi_k(L, K) \xrightarrow{a_*} \pi_k(\varphi) \xrightarrow{b_*} \pi_k(\Psi) \xrightarrow{\partial} \pi_{k-1}(L, K) \xrightarrow{a_*} \cdots \tag{$*$}$$

Since $L = K \cup_{f_j} \{e_j^n\}_{j \in J}$ we have $\pi_k(L, K) = 0$ for $k \le n - 1$ and $\pi_n(L, K)$ is generated over $\pi_1(K)$ by the classes $[\chi_j]$ where $\chi_j: (D^n, S^{n-1}) \to (L, K)$ is the characteristic map for the n cell e_j^n. Clearly $a_*([\chi_j]) = \alpha_j \in \pi_n(\varphi)$. Thus $a_*: \pi_n(L, K) \to \pi_n(\varphi)$ is onto. It follows from the exactness of $(*)$ that $\pi_n(\Psi) = 0$ and that $0 = \pi_k(\varphi) \simeq \pi_k(\Psi)$ for $k \le n - 1$. In other words Ψ is n-connected.

Proposition 1.4

Let K be a CW-complex, X of the homotopy type of a CW-complex, and $\varphi: K \to X$ an $(n - 1)$-connected map with $n \ge 2$. Then there exists a CW-complex $\Gamma = K \cup (\text{cells of dim} \ge n)$ and a homotopy equivalence (abbreviated as h.e.) $\theta: \Gamma \to X$ extending φ.

Proof. By applying Lemma 1.3 repeatedly we get CW-complexes K_ℓ and maps $\varphi_\ell: K_\ell \to X$ for $\ell \ge n$ satisfying

(i) $K_n = K \cup (\text{cells of dim } n)$.
(ii) $K_{\ell+1} = K_\ell \cup (\text{cells of dim } \ell + 1)$ for $\ell \ge n$.
(iii) $\varphi_n | K = \varphi$.
(iv) $\varphi_{\ell+1} | K_\ell = \varphi_\ell$ for $\ell \ge n$.
(v) φ_ℓ is ℓ-connected for $\ell \ge n$.

If $\Gamma = \bigcup_{\ell \ge n} K_\ell$, then there is a well-defined map $\theta: \Gamma \to X$ given by $\theta | K_\ell = \varphi_\ell$. Also θ is ℓ-connected for all ℓ. Hence by Whitehead's theorem, θ is a h.e. $\square$

Proposition 1.5

Let $n \ge 3$, $f: K = K^{n-1} \to X = X^n$ be an $(n - 1)$-connected map with K and X finite complexes. Then $\pi_n(f)$ is a finitely generated projective $Z\pi$-module where $\pi = \pi_1(X)$.

Proof. Observe that since $n \ge 3$ and f is $(n - 1)$-connected we have $f_*: \pi_1(K) \simeq \pi_1(X) = \pi$. Hence if $\tilde{f}: \tilde{K} \to \tilde{X}$ denotes the lift to universal coverings, we have $\pi_n(\tilde{f}) \simeq \pi_n(f)$ and by Hurewicz theorem $\pi_n(\tilde{f}) \simeq H_n(\tilde{f})$. Observe that the mapping cylinder M_f of f is also a finite CW-complex of dimension less than or equal to n and we can replace f by the inclusion of K into M_f up to homotopy. Thus without loss of generality we can assume that f is the inclusion j of a subcomplex $K = K^{n-1}$ into $X = X^n$ with X finite. Then j is $(n-1)$-connected and $\pi_n(j) \simeq \pi_n(\tilde{j}) \simeq H_n(\tilde{j}) \simeq H_n(\tilde{X}, \tilde{K})$. Since $(\tilde{X}, \tilde{K})$ is a relative CW-complex, with $\dim \tilde{X} \le n$, the cellular chain complex of $(\tilde{X}, \tilde{K})$ is of the form

$$\cdots 0 \to 0 \to C_n(\tilde{X}, \tilde{K}) \to C_{n-1}(\tilde{X}, \tilde{K}) \to \cdots$$

$$\cdots \to \cdots \to C_1(\tilde{X}, \tilde{K}) \to C_0(\tilde{X}, \tilde{K}) \to 0 \to 0 \to \cdots.$$

Moreover, since X is a finite CW-complex each $C_k(\tilde{X}, \tilde{K})$ is a free finitely generated $Z\pi$-module.

Let

$$\left. \begin{array}{l} Z_k(\tilde{X}, \tilde{K}) = \ker d_k \colon C_k(\tilde{X}, \tilde{K}) \to C_{k-1}(\tilde{X}, \tilde{K}) \\[2mm] B_k(\tilde{X}, \tilde{K}) = \operatorname{Im} d_{k+1} \colon C_{k+1}(\tilde{X}, \tilde{K}) \to C_k(\tilde{X}, \tilde{K}) \end{array} \right\} \quad \text{for } k \ge 0.$$

Then $C_0(\tilde{X}, \tilde{K}) = Z_0(\tilde{X}, \tilde{K})$, $Z_k(\tilde{X}, \tilde{K}) = B_k(\tilde{X}, \tilde{K})$ for $k \le n-1$, and $H_n(\tilde{X}, \tilde{K}) = Z_n(\tilde{X}, \tilde{K})$.

We thus have the following exact sequences of $Z\pi$-modules:

$$0 \to Z_1(\tilde{X}, \tilde{K}) \to C_1(\tilde{X}, \tilde{K}) \to C_0(\tilde{X}, \tilde{K}) \to 0, \tag{1}$$

$$0 \to Z_2(\tilde{X}, \tilde{K}) \to C_2(\tilde{X}, \tilde{K}) \to Z_1(\tilde{X}, \tilde{K}) \to 0, \tag{2}$$

$$\cdots$$

$$0 \to H_n(\tilde{X}, \tilde{K}) \to C_n(\tilde{X}, \tilde{K}) \to Z_{n-1}(\tilde{X}, \tilde{K}) \to 0. \tag{n}$$

Since $C_0(\tilde{X}, \tilde{K})$ is free $Z\pi$, sequence (1) splits showing that $Z_1(\tilde{X}, \tilde{K})$ is projective over $Z\pi$. This implies that sequence (2) splits and hence $Z_2(\tilde{X}, \tilde{K})$ is a direct summand of $C_2(\tilde{X}, \tilde{K})$ and hence projective over $Z\pi$. Proceeding thus we finally see that $H_n(\tilde{X}, \tilde{K})$ is a direct summand of $C_n(\tilde{X}, \tilde{K})$ and hence $H_n(\tilde{X}, \tilde{K})$ is finitely generated projective over $Z\pi$. $\square$

Proposition 1.6

Let $\varphi \colon K = K^2 \to X = X^2$ be a map inducing an isomorphism of fundamental groups. Let K and X be finite complexes. Then $\pi_2(\varphi)$ is finitely generated over $Z\pi$ where $\pi = \pi_1(X)$.

Proof. Let $Y = M_\varphi$ and $j: K \to M_\varphi$ denote the inclusion of K into M_φ. Then Y is a finite complex with $\dim Y \le 3$ and $j: K \to Y$ induces isomorphism of fundamental groups. Up to homotopy φ could be replaced by j. Since $j_*: \pi_1(K) \simeq \pi_1(Y)$ we see that $\pi_1(j) = 0$. Hence $\pi_2(j) \simeq \pi_2(\tilde{j}) \simeq H_2(\tilde{j})$ where $\tilde{j}: \tilde{K} \to \tilde{Y}$ is the lift of j to universal coverings. The cellular complex of $(\tilde{Y}, \tilde{K})$ is of the form

$$\cdots \to 0 \to 0 \to C_3(\tilde{Y}, \tilde{K}) \to C_2(\tilde{Y}, \tilde{K})$$

$$\to C_1(\tilde{Y}, \tilde{K}) \to C_0(\tilde{Y}, \tilde{K}) \to 0 \to 0 \cdots.$$

From $H_0(\tilde{Y}, \tilde{K}) = 0 = H_1(\tilde{Y}, \tilde{K})$ we get exact sequences

$$0 \to Z_1(\tilde{Y}, \tilde{K}) \to C_1(\tilde{Y}, \tilde{K}) \to C_0(\tilde{Y}, \tilde{K}) \to 0, \tag{1}$$

$$0 \to Z_2(\tilde{Y}, \tilde{K}) \to C_2(\tilde{Y}, \tilde{K}) \to Z_1(\tilde{Y}, \tilde{K}) \to 0, \tag{2}$$

where $Z_k(\tilde{Y}, \tilde{K}) = \ker d_k: C_k(\tilde{Y}, \tilde{K}) \to C_{k-1}(\tilde{Y}, \tilde{K})$. Since $C_0(\tilde{Y}, \tilde{K})$ is free over $Z\pi$, sequence (1) splits showing that $Z_1(\tilde{Y}, \tilde{K})$ is projective over $Z\pi$. Hence sequence (2) also splits.

It follows that $Z_2(\tilde{Y}, \tilde{K})$ is a direct summand of $C_2(\tilde{Y}, \tilde{K})$ and hence finitely generated projective over $Z\pi$. Now $H_2(\tilde{Y}, \tilde{K})$ is a quotient of $Z_2(\tilde{Y}, \tilde{K})$ and hence $H_2(\tilde{Y}, \tilde{K})$ is finitely generated over $Z\pi$. $\square$

Let X denote a 0-connected space of the homotopy type of a CW-complex. In what follows π will denote $\pi_1(X)$. We will describe certain conditions on X that will ensure that X is h.e. to a CW-complex with K^n finite, where n is a given integer greater than or equal to 1. However these conditions have to be broken into three "groups" depending on whether $n = 1, 2,$ or ≥ 3.

Definition 1.7

(i) We say that X satisfies F_1 if π is finitely generated.

(ii) We say that X satisfies F_2 if π is finitely presented and for any finite complex $K = K^2$ and any map $\varphi: K^2 \to X$ inducing an isomorphism of fundamental groups, $\pi_2(\varphi)$ is f.g. (finitely generated) over $Z\pi$.

(iii) Let $n \ge 3$. We say that X satisfies F_n if it satisfies F_{n-1} and for any $(n-1)$-connected map $\varphi: K = K^{n-1} \to X$ with K finite, $\pi_n(\varphi)$ is f.g. over $Z\pi$.

Definition 1.8

We say that X satisfies F_∞ if X satisfies F_n for all $n \ge 1$.

Before dealing with the main result, we state and prove an algebraic result due to J. R. Stallings.

Definition 1.9

A group H will be called a retract of another group G if $\exists$ homomorphisms $j\colon H \to G$ and $\varepsilon\colon G \to H$ with $\varepsilon \cdot j = \mathrm{Id}_H$. In this case clearly j is an injection and ε a surjection. We will refer to ε as a retraction of G onto H.

Proposition 1.10 (Stallings)

Let G be a finitely presented group and H a retract of G. Let $j\colon H \to G$ and $\varepsilon\colon G \to H$ be homomorphisms satisfying $\varepsilon j = \mathrm{Id}_H$. Let $\langle x_1, \ldots, x_k; r_1, \ldots, r_\ell \rangle$ be a presentation of G. Then there exists a presentation $\langle x_1, \ldots, x_k, r_1, \ldots, r_\ell, u_1, \ldots, u_k \rangle$ for H such that ε corresponds to the canonical quotient map

$$\langle x_1, \ldots, x_k; r_1, \ldots, r_\ell \rangle \to \langle x_1, \ldots, x_k; r_1, \ldots, r_\ell, u_1, \ldots, u_k \rangle.$$

Proof. Let F be the free group on $x_1, \ldots, x_k$ and N the *normal* subgroup of F generated by $r_1, \ldots, r_\ell$. Then G can be identified with F/N. Let $\eta\colon F \to G/N = G$ denote the canonical quotient map. Then $j\varepsilon\eta(x_i) \in G$. Hence there exists an element $w_i \in F$ with $\eta(w_i) = j\varepsilon\eta(x_i)$ in F/N. Let $u_i = x_i^{-1}w_i \in F$ $(1 \le i \le k)$ and Γ be the normal subgroup of F generated by $r_1, \ldots, r_\ell, u_1, \ldots, u_k$. Since $\Gamma \supset N$, the canonical quotient maps $F \overset{\xi}{\to} F/\Gamma$, $F/N \overset{\beta}{\to} F/\Gamma$ make

$$
\begin{array}{ccc}
 & F & \\
{\scriptstyle \eta}\swarrow & & \searrow{\scriptstyle \xi} \\
F/N & \underset{\beta}{\longrightarrow} & F/\Gamma
\end{array}
$$

commutative. Now,

$$
\begin{aligned}
\varepsilon\eta(u_i) &= \varepsilon\eta\big(x_i^{-1}\big)\varepsilon\eta(w_i) \\
&= \varepsilon\eta\big(x_i^{-1}\big)\varepsilon j\varepsilon\eta(x_i) \\
&= \varepsilon\eta\big(x_i^{-1}\big)\varepsilon\eta(x_i) \quad \text{since } \varepsilon j = \mathrm{Id}_H \\
&= \varepsilon\eta\big(x_i^{-1} \cdot x_i\big) = 1 \quad \text{in } H \text{ for } 1 \le i \le k.
\end{aligned}
$$

Also $\varepsilon\eta(r_\mu) = \varepsilon(1)$ in H for $1 \le \mu \le \ell$.

It follows that $\exists$ a homomorphism $F/\Gamma \overset{\theta}{\to} H$ with

Diagram B.

commutative. To prove Proposition 1.10, we have only to show that $\theta: F/\Gamma \to H$ is an isomorphism. Then if we identify H with F/Γ using the isomorphism $\theta: F/\Gamma \to H$, the map $\varepsilon: F/N = G \to H$ gets identified with the canonical quotient map $\beta: F/N \to F/\Gamma$.

Let $\alpha = \beta j: H \to F/\Gamma$. Then $\theta\alpha = \theta\beta j = \varepsilon j = \mathrm{Id}_H$. This in particular shows that

$$\alpha: H \to F/\Gamma \text{ is monic} \quad \text{and} \quad \theta: F/\Gamma \to H \text{ is epic.} \tag{i}$$

Also $\varepsilon\eta(x_i) \in H$ satisfies

$$\alpha\varepsilon\eta(x_i) = \beta j\varepsilon\eta(x_i) = \beta\eta(w_i)$$

$$= \zeta(w_i)$$

$$= \zeta(x_i)\zeta(x_i^{-1}w_i)$$

$$= \zeta(x_i) \cdot 1 = \zeta(x_i).$$

We know that F/Γ is generated by $\zeta(x_i)$, $1 \le i \le k$. It follows that $\alpha: H \to F/\Gamma$ is onto. Combining this with the already established fact that α is monic we get $\alpha: H \simeq F/\Gamma$. From $\theta\alpha = \mathrm{Id}_H$ we see that $\theta: F/\Gamma \to H$ is the inverse of α. This completes the proof of Proposition 1.10. $\square$

This result of Stallings appears in [24].

Theorem 1.11

Let n be an integer greater than or equal to 1. The following conditions on X are equivalent:

(i) *X is homotopically equivalent to a CW-complex with finite n skeleton.*

(ii) *X is dominated by a CW-complex with finite n skeleton.*

(iii) *X satisfies F_n.*

Proof. (i) $\Rightarrow$ (ii) is trivial. We will prove that (i) $\Leftrightarrow$ (iii) and that (ii) $\Rightarrow$ (i). This proves the theorem.

(i) $\Rightarrow$ (iii). We may assume that X is itself a CW-complex with X^n finite. If $n = 1$, then $\pi_1(X) \simeq \pi_1(X^2)$ is finitely presented. Let $\varphi: K = K^2 \to X$ be any map with $\varphi_*: \pi_1(K) \simeq \pi_1(X)$ and K finite. By the cellular approximation theorem we may assume $\varphi(K^2) \subset X^2$. Hence $\varphi = i \circ \theta$ with $\theta: K^2 \to X^2$ and $i: X^2 \to X$. Then $\theta_*: \pi_1(K) \simeq \pi_1(X^2)$. If $\tilde{K}^2 \overset{\tilde{\theta}}{\to} \tilde{X}^2 \overset{\tilde{i}}{\to} \tilde{X}$ denote lifts to universal coverings, then $\tilde{\varphi} = \tilde{i} \circ \tilde{\theta}$ lifts φ and $\pi_2(\varphi) \simeq \pi_2(\tilde{\varphi}) \simeq H_2(\tilde{\varphi})$.

The exactness of $H_2(\tilde{\theta}) \to H_2(\tilde{\varphi}) \to H_2(\tilde{X}, \tilde{X}^2)$ together with $H_2(\tilde{X}, \tilde{X}^2) = 0$ shows that $H_2(\tilde{\varphi})$ is a quotient of $H_2(\tilde{\theta})$. Proposition 1.6 shows that $\pi_2(\tilde{\theta}) \simeq \pi_2(\theta) \simeq H_2(\tilde{\theta})$ is f.g. over $Z\pi$. Hence $\pi_2(\varphi) \simeq H_2(\tilde{\varphi})$ is f.g. over $Z\pi$. This proves that X satisfies F_2.

Suppose $n \geq 3$. By induction, we may assume X satisfies F_{n-1}. Let $\varphi: K = K^{n-1} \to X$ be any $(n - 1)$-connected map with K finite. By the cellular approximation theorem we can assume $\varphi(K^{n-1}) \subset X^{n-1} \subset X^n$. Thus $\varphi = i \circ \theta$ with $\theta: K^{n-1} \to X$ and $i: X^n \to X$. Now the map θ is $(n-1)$-connected, since $i_*: \pi_j(X^n) \simeq \pi_j(X)$ for $j \leq n - 1$. By Proposition 1.5, $\pi_n(\theta)$ is f.g. projective over $Z\pi$. Also using lifts $\tilde{K} \overset{\tilde{\theta}}{\to} \tilde{X}^n \overset{\tilde{i}}{\to} \tilde{X}$ of θ and i and taking $\tilde{\varphi} = \tilde{i} \circ \tilde{\theta}$ we see that $H_n(\tilde{\theta}) \to H_n(\tilde{\varphi}) \to H_n(\tilde{X}, \tilde{X}^n)$ is exact. Thus $\pi_n(\varphi) \simeq \pi_n(\tilde{\varphi}) \simeq H_n(\tilde{\varphi})$ is a quotient of $H_n(\tilde{\theta})$, since $H_n(\tilde{X}, \tilde{X}^n) = 0$. Hence $\pi_n(\varphi)$ is f.g. over $Z\pi$. This shows that X satisfies F_n.

(iii) $\Rightarrow$ (i). Let X satisfy F_1, i.e., $\pi_1(X)$ is f.g. Then there exists a finite wedge of circles $K = K^1 = \bigvee_{i=1}^{k} S_i^1$ and a map $\varphi: K \to X$ that induces an onto map $\varphi_*: \pi_1(K) \to \pi_1(X)$. Hence $\pi_1(\varphi) = 0$. The arc-connected nature of K and X yields $\pi_0(\varphi) = 0$. Thus φ is 1-connected. From Proposition 1.4 we get a CW-complex $\Gamma = K \cup$ (cells of dim ≥ 2) and a h.e. $\theta: \Gamma \to X$ extending φ. Now $\Gamma^1 = K^1$ is finite. Thus X is homotopically equivalent to a complex Γ with Γ^1 finite.

Suppose X satisfies F_2. Then $\pi_1(X)$ is finitely presented. Hence $\exists$ a finite $K = K^2$ and a map $\varphi: K \to X$ inducing an isomorphism $\varphi_*: \pi_1(K) \simeq \pi_1(X)$. By F_2, $\pi_2(\varphi)$ is a f.g. $Z\pi$-module. By the construction of Milnor and Lemma 1.3, $\exists$ a complex $K_2 = K \cup$ (finitely many two cells) and a 2-connected map $\varphi_2: K_2 \to X$ extending φ. Now, by Proposition 1.4, $\exists$ a CW-complex $\Gamma = K_2 \cup$ (cells of dim ≥ 3) and a h.e. $\theta: \Gamma \to X$ extending φ_2. Now, $\Gamma^2 = K_2^2 = K^2 \cup$ (finite number of 2 cells) is finite. Thus X is homotopically equivalent to Γ where Γ^2 is finite.

For $n \geq 2$, by induction on n, we will prove that if X satisfies F_n then $\exists$ an n-connected map $K = K^n \overset{\varphi}{\to} X$ with K finite. For $n = 2$, we verified the validity of this statement in the preceding paragraph, namely when X satisfied F_2, we constructed a 2-connected map $K_2 \overset{\varphi_2}{\longrightarrow} X$ with $K_2 = K_2^2$ finite. Let $n \geq 3$ and let X satisfy F_n. Then by definition X satisfies F_{n-1}. Hence, by the inductive assumption $\exists$ a finite complex $K = K^{n-1}$ and an $(n-1)$-connected map $\alpha: K^{n-1} \to X$. By F_n, $\pi_n(\alpha)$ is f.g. over $Z\pi$. By applying Milnor's construction, we see from Lemma 1.3 that $\exists$ an n-connected map $\varphi: K_n \to X$ extending α where $K_n = K^{n-1} \cup$ (finitely many n cells). Clearly $K_n = K_n^n$ is finite. This completes the inductive step.

Thus if $n \geq 3$ and X satisfies F_n, we get a finite complex $K = K^n$ and an n-connected map $\varphi: K \to X$. From Proposition 1.4 we get a complex $\Gamma = K \cup$ (cells of dim $\geq n+1$) and a h.e. $\theta: \Gamma \to X$ extending φ. Now $\Gamma^n = K^n$ is finite. This completes the proof of (iii) $\Rightarrow$ (i).

(ii) $\Rightarrow$ (i). Let $X \xrightarrow{f} L \xrightarrow{g} X$ satisfy $gf \sim \text{Id}_X$ where L is a CW-complex with L^n finite. Since $gf \sim \text{Id}_X$, $(gf)_*: \pi_1(X, x_0) \simeq \pi_1(X, gf(x_0))$. If a is the inverse isomorphism $\pi_1(xfg(x_0)) \to \pi_1(X, x_0)$ to $(gf)_*$, then $f_* \circ a: \pi_1(X, gf(x_0)) \to \pi_1(L, f(a_0))$ and $g_*: \pi_1(L, f(a_0)) \to \pi_1(X, fg(x_0))$ satisfy $g_* \circ f_* \circ a = \text{Id}_{\pi_1(x, gf(x_0))}$.

Choosing $f(x_0)$ as base point in L and $gf(x_0)$ as base point in the second copy of X, we see that $g_*: \pi_1(L) \to \pi_1(X)$ realizes $\pi_1(X)$ as a retract of $\pi_1(L)$.

Let L^n be finite. If $n = 1$, $\pi_1(L)$ is f.g. and hence $\pi_1(X)$ is f.g. In particular X satisfies F_1. The proof of (iii) $\Rightarrow$ (i) for $n = 1$, shows that X is homotopically equivalent to a complex Γ with Γ^1 finite.

Let $n = 2$. The $\pi_1(L)$ is finitely presented. Since $g_*: \pi_1(L) \to \pi_1(X)$ is a retraction, by Proposition 1.10 $\pi_1(X)$ is also finitely presented. In fact if $\langle x_1, \ldots, x_k; r_1, \ldots, r_\ell \rangle$ is a presentation for $\pi_1(L)$ and $\eta: F \to \pi_1(L)$ the quotient map where F equals the free group on $x_1, \ldots, x_k$, then the proof of Proposition 1.10 shows that $\exists$ element $u_1, \ldots, u_k$ in F with $\langle s_1, \ldots, x_k; r_1, \ldots, r_\ell, u_1, \ldots, u_k \rangle$ yielding a presentation of $\pi_1(X)$ in such a way that g_* corresponds to the obvious quotient map $\langle x_1, \ldots, x_k; r_1, \ldots, r_\ell \rangle \to \langle x_1, \ldots, x_k; r_1, \ldots, r_\ell, u_1, \ldots, u_k \rangle$. If $c_i = \eta(u_i) \in \pi_1(L)$, then $g_*: \pi_1(L) \to \pi_1(X)$ precisely kills c_i, $1 \leq i \leq k$. Let $\mu_i = S^1 \to L$ represent c_i. Then $\exists$ an extension $g_2: L_2 = L \cup_{\mu_i} \{e_i^2\}_{i=1}^k \to X$ extending g and $g_{2*}: \pi_1(L_2) \simeq \pi_1(X)$. Also $g_2 f = gf \sim \text{Id}_X$. From the exact sequence

$$\pi_2(L_2) \xrightarrow{g_{2*}} \pi_2(X) \longrightarrow \pi_2(g_2) \longrightarrow \pi_1(L_2) \xrightarrow[\simeq]{g_{2*}} \pi_1(X)$$

$$\longrightarrow \pi_1(g_2) \longrightarrow \cdots$$

we see that $\pi_0(g_2) = \pi_1(g_2) = 0$. Also $g_2 f \sim \text{Id}_X \Rightarrow \pi_2(L_2) \xrightarrow{g_{2*}} \pi_2(X)$. Hence $\pi_2(g_2) = 0$. Thus $g_2: L_2 \to X$ is a 2-connected map and $L_2^2 = L^2 \cup \{e_i^2\}_{i=1}^k$ if finite. By Proposition 1.4, $\exists$ a CW-complex $\Gamma = L_2 \cup$ (cells of dim ≥ 3) and a h.e. $\theta: \Gamma \to X$ extending g_2. Now $\Gamma^2 = L_2^2$ is finite.

Let $X \xrightarrow{f} L \xrightarrow{g} X$ satisfy $gf \sim \text{Id}_X$ where L is a CW-complex with L^n finite. For $n \geq 2$, by induction on n we will prove the following statement: $\exists$ a complex $L_n = L \cup \{$finitely many cells in dim $\leq n\}$ and an n-connected map

$g_n: L_n \to X$ extending g. For $n = 2$ this assertion was just now proved. Let $n \geq 3$. By the inductive assumption there exists an $L_{n-1} = L \cup$ (finitely many cells in dim $\leq n - 1$) and an $(n - 1)$-connected map $g_{n-1}: L_{n-1} \to X$ extending g. Write L' for L_{n-1} and g' for $g_{n-1}: L_{n-1} \to X$. By Proposition 1.4 there exists a CW-complex $\Gamma = L' \cup$ (cells in dim $\geq n$, possibly infinitely many in certain dimensions) and a h.e. $\theta: \Gamma \to X$ extending g'. Let $i: L' \to \Gamma$ denote the inclusion. Let $\mu: \Gamma \to L'$ be the map $\mu = f\theta$. Then

$$
\begin{array}{ccc}
L' & \xrightarrow{\ g'\ } & X \\
{\scriptstyle i}\big\downarrow & & \big\| \\
\Gamma & \xrightarrow{\ \theta\ } & X
\end{array}
$$

Diagram C.

is commutative.

Also $\theta i\mu = \theta i f\theta - g'f\theta = gf\theta \sim \mathrm{Id}_X \circ \theta = \theta = \theta \circ \mathrm{Id}_\Gamma$. Thus $\theta \circ i\mu \sim \theta \circ \mathrm{Id}_\Gamma$. Since θ is a h.e. we see that $i\mu \sim \mathrm{Id}_\Gamma$. In other words the composite

$$
\Gamma \xrightarrow{\ \mu\ } L' \xrightarrow{\ i\ } \Gamma
$$

is homotopic to Id_Γ. Moreover $\Gamma^{n-1} = L'^{n-1}$ and $i: L' \to \Gamma$ is $(n-1)$-connected, since $g': L' \to X$ is $(n-1)$-connected. In particular $i_*: \pi_1(L') \simeq \pi_1(\Gamma)$. Let

$$
\tilde{\Gamma} \xrightarrow{\ \tilde{\mu}\ } \tilde{L}' \xrightarrow{\ \tilde{i}\ } \tilde{\Gamma}
$$

denote lifts to universal coverings. Then $\tilde{i}\tilde{\mu} \sim \mathrm{Id}_{\tilde{\Gamma}}$. Hence for each j, the composite $\tilde{i}_* \circ \tilde{\mu}_*$ in

$$
H_j(\tilde{\Gamma}) \xrightarrow{\ \tilde{\mu}_*\ } H_j(\tilde{L}') \xrightarrow{\ \tilde{i}_*\ } H_j(\tilde{\Gamma})
$$

is $\mathrm{Id}_{H_j(\tilde{\Gamma})}$. In particular $\tilde{\mu}_*: H_j(\tilde{\Gamma}) \to H_j(\tilde{L}')$ is monic and $\tilde{\mu}_*(H_j(\tilde{\Gamma}))$ is a direct summand of $H_j(\tilde{L}')$ (as a $Z\pi$-module).

The exact homology sequence

$$
H_{j+1}(\tilde{\mu}) \xrightarrow{\ \partial\ } H_j(\tilde{\Gamma}) \overset{\tilde{\mu}_*}{\rightarrowtail} H_j(\tilde{L}') \longrightarrow H_j(\tilde{\mu}) \xrightarrow{\ \partial\ } H_{j-1}(\tilde{\Gamma}) \overset{\tilde{\mu}_*}{\rightarrowtail} H_{j-1}(\tilde{L}') \longrightarrow
$$

for the map $\tilde{\mu}$, shows that

$$
H_j(\tilde{L}) = \tilde{\mu}_*\big(H_j(\tilde{\Gamma})\big) \oplus A_j \quad \text{with } A_j \simeq H_j(\tilde{\mu}) \text{ in } Z\pi\text{-mod.} \tag{1}
$$

Similarly the exact homology sequence of the map $\tilde{i}$,

$$
H_{j+1}(\tilde{L}') \underset{\tilde{\mu}_* \text{ splits } \tilde{i}_*}{\overset{\tilde{i}_*}{\rightleftarrows}} H_{j+1}(\tilde{\Gamma}) \longrightarrow H_{j+1}(\tilde{i}) \longrightarrow H_j(\tilde{L}') \underset{\tilde{\mu}_* \text{ splits } \tilde{i}_*}{\overset{\tilde{i}_*}{\rightleftarrows}} H_j(\tilde{\Gamma}) \longrightarrow
$$

we see that

$$H_j(\tilde{L}') = \tilde{\mu}_*\big(H_j(\tilde{\Gamma})\big) \oplus B_j \quad \text{with } B_j \simeq H_{j+1}(\tilde{i}) \text{ in } Z\pi\text{-mod.} \qquad (2)$$

From (1) and (2) we see that $A_j \simeq B_j$ in $Z\pi$-mod. In particular

$$\boxed{H_{n-1}(\tilde{\mu}) \simeq H_n(\tilde{i}) \quad \text{in } Z\pi\text{-mod.}} \qquad (3)$$

Let $k = \mu|\Gamma^{n-2}\colon \Gamma^{n-2} \to L'$ and let $\tilde{\Gamma}^{n-2}$ be the inverse image of Γ^{n-2} under the projection map $\tilde{\Gamma} \to \Gamma$. (If $n \geq 4$, $\tilde{\Gamma}^{n-2}$ will be the universal cover of Γ^{n-2}.) Let $\tilde{k} = \tilde{\mu}|\tilde{\Gamma}^{n-2}\colon \tilde{\Gamma}^{n-2} \to \tilde{L}'$. The exactness of

$$H_{n-1}(\tilde{\Gamma}, \tilde{\Gamma}^{n-2}) \longrightarrow H_{n-1}(\tilde{k}) \longrightarrow H_{n-1}(\tilde{\mu}) \longrightarrow \overset{\overset{\displaystyle 0}{\|\|}}{H_{n-2}(\tilde{\Gamma}, \tilde{\Gamma}^{n-2})} \qquad (4)$$

shows that $H_{n-1}(\tilde{\mu})$ is a quotient of $H_{n-1}(\tilde{k})$.

We know that $i \circ \mu \sim \mathrm{Id}_\Gamma$. Without loss of generality we can assume μ to be cellular, hence $\mu(\Gamma^{n-2}) \subset L'^{n-2} = \Gamma^{n-2} \subset \Gamma^{n-1}$. $i \circ \mu \sim \mathrm{Id}_\Gamma \Rightarrow i \circ k \sim \nu_{n-2}\colon \Gamma^{n-2} \to \Gamma$ where ν_{n-2} is the inclusion $\Gamma^{n-2} \to \Gamma$. Both $i \circ k$ and ν_{n-2} are cellular. By the cellular homotopy theorem, we can assume that $\exists$ a homotopy between $i \circ k$ and ν_{n-2} taking place in Γ^{n-1}. Since $L'^{n-1} = \Gamma^{n-1}$, the map $i \circ k$ thought of as a map of Γ^{n-2} into Γ^{n-1} is the same as k as a map of Γ^{n-2} into Γ^{n-1}. Thus k regarded as a map of Γ^{n-2} in Γ^{n-1} is homotopic to the inclusion $\Gamma^{n-2} \to \Gamma^{n-1}$. Using the fact that $\Gamma^{n-2} \to \Gamma$ is a cofibration we can assume that $\mu\colon \Gamma \to L'$ itself satisfies the condition that $k = \mu|\Gamma^{n-2}$ is the inclusion of $\Gamma^{n-2} = L'^{n-2}$ into L'.

The exact sequence

$$H_{n-1}(\tilde{L}'^{n-1}, \tilde{L}'^{n-2}) \longrightarrow H_{n-1}(\tilde{k}) \longrightarrow \overset{\overset{\displaystyle 0}{\|\|}}{H_{n-1}(\tilde{L}', \tilde{L}'^{n-1})}$$

shows that $H_{n-1}(\tilde{k})$ is a quotient of $H_{n-1}(\tilde{L}'^{n-1}, \tilde{L}'^{n-2})$. But L'^{n-1} is finite. Hence $H_{n-1}(\tilde{L}'^{n-1}, \tilde{L}'^{n-2})$ is f.g. over $Z\pi$. It follows that $H_{n-1}(\tilde{k})$ is f.g. over $Z\pi$. Using (4) and (3) we see that $H_n(\tilde{i})$ is f.g. over $Z\pi$.

By Proposition 1.4, $\exists$ a complex $L_n = L' \cup$ (finitely many n cells) and a map $\gamma\colon L_n \to \Gamma$ extending $i\colon L' \to \Gamma$ with γ n-connected. Then $\theta \circ \gamma\colon L_n \to X$ is an n-connected map extending $g_{n-1}\colon L_{n-1} \to X$. Moreover $L_n = L_{n-1} \cup$ (finitely many n cells). Hence L_n^n is finite. The completes the inductive step.

In fact, in the inductive construction, we got L_n as $L_{n-1} \cup$ (finitely many n cells) and an n-connected map $g_n: L_n \to X$ extending g_{n-1}. To assert that L_n^n is finite, we need to know that L^n is finite.

By Proposition 1.4 $\exists$ a complex $E = L_n \cup$ (cells in dim $\geq n + 1$) and a h.e. $\delta: E \to X$ extending g_n. Now $E^n = L_n^n$ is finite. This proves (ii) $\Rightarrow$ (i). $\square$

Theorem 1.12

The following conditions on X are equivalent:

  (i) *X is homotopically equivalent to a CW-complex of finite type.*

  (ii) *X is dominated by a CW-complex of finite type.*

  (iii) *X satisfies F.*

Proof. (i) $\Rightarrow$ (iii). We may assume that X itself is a CW-complex of finite type. Then (i) $\Rightarrow$ (iii) of Theorem 1.11 shows that X satisfies F_n for all n. Hence X satisfies F.

(iii) $\Rightarrow$ (i). In fact the proof of (iii) $\Rightarrow$ (i) in Theorem 1.11 yields *finite* complexes K_n for $n \geq 3$ and n-connected maps $\varphi_n: K_n \to X$ satisfying

  (a) $K_{n+1} = K_n \cup$ [finitely many $(n + 1)$ cells].

  (b) $K_n^n = K_n$.

  (c) $\varphi_{n+1}|K_n = \varphi_n$.

If $K = \bigcup_{n \geq 3} K_n$, then $\varphi: K \to X$ defined by $\varphi|K_n = \varphi_n$ (for $n \geq 3$) is a h.e. Clearly K is of finite type.

(i) $\Rightarrow$ (ii). Trivial.

(ii) $\Rightarrow$ (i). Let $X \overset{f}{\to} L \overset{g}{\to} X$ be such that $gf \sim \mathrm{Id}_X$ and L a CW-complex of finite type. The proof of (ii) $\Rightarrow$ (i) of Theorem 1.11 yields complexes L_n and maps $g_n: L_n \to X$ for $n \geq 2$ satisfying

  (a) $L_2 = L \cup$ (finitely many 2 cells).

  (b) $L_{n+1} = L_n \cup$ [finitely many $(n + 1)$ cells] (for $n \geq 2$).

  (c) $g_2|L = g$ and $h_{n+1}|L_n = g_n$ for $n \geq 2$.

  (d) g_n is n-connected.

Let $E = \bigcup_{n \geq 2} L_n$ and $\varphi: E \to X$ be given by $\varphi|L_n = g_n$. Then E is a CW-complex of finite type and φ is a h.e.

Remark 1.13

Given X and any integer $n \geq 2$, there exists an n-connected map $f: K \to X$ with K a CW-complex of dimension less than or equal to n.

Proof. Definitely we can get a map $\varphi\colon A = A^2 \to X$ with $\varphi_*\colon \pi_1(A) \simeq \pi_1(X)$. In this case φ is automatically 1-connected. (i) now follows from a repeated application of Milnor's construction. $\square$

Remark 1.14

If $\pi = \pi_1(X)$ is finitely presented, $\exists$ a *finite* complex $K = K^2$ and a map $\varphi\colon K \to X$ with $\varphi_*\colon \pi_1(K) \simeq \pi_1(X)$.

Remark 1.15

Suppose X satisfies F_n with $n \geq 2$. While proving (iii) $\Rightarrow$ (i) of Theorem 1.11 we have shown that there exists an n-connected map $\varphi\colon K = K^n \to X$ with K finite.

Remark 1.16

Let $n \geq 2$ and X satisfy F_n. Then as seen already $\exists$ an n-connected map $\varphi\colon K = K^n \to X$ with K finite. Suppose $\pi_{n+1}(\varphi)$ is f.g. over $Z\pi$ [where $\pi = \pi_1(X)$]. Then by Milnor's construction, $\exists$ an $(n+1)$-connected map $\varphi_{n+1}\colon K_{n+1} \to X$ extending φ where $K_{n+1} = K \cup$ [finitely many $(n+1)$ cells]. Again by Proposition 1.4 φ_{n+1} extends to a h.e. $\theta\colon \Gamma \to X$ with $\Gamma = K_{n+1} \cup$ (cells of dim $\geq n+1$). Clearly $\Gamma^{n+1} = K_{n+1}$ is finite. From (i) $\Rightarrow$ (iii) of Theorem 1.11 we see that X satisfies F_{n+1}. Hence for every n-connected map $f\colon L \to X$ with $L = L^n$ finite we will have $\pi_{n+1}(f)$ f.g. over $Z\pi$.

Remark 1.17

Let π be finitely presented. Then $\exists$ a $\varphi\colon K = K^2 \to X$ with $\varphi_*\colon \pi_1(K) \simeq \pi_1(X)$ as in 1.14. If $\pi_2(\varphi)$ is f.g. over $Z\pi$ we can extend φ to a 2-connected map $\varphi_2\colon K_2 = K \cup$ (finitely many 2 cells) $\to X$. Proposition 1.4 now yields a h.e. $\theta\colon \Gamma \to X$ extending φ_2 with $\Gamma = K_2 \cup$ (cells of dim ≥ 3). Clearly $\Gamma^2 = K_2$ is finite. By (i) $\Rightarrow$ (iii) of Theorem 1.11, X satisfies F_2. Hence for any $f\colon L = L^2 \to X$ with L finite and $f_*\colon \pi_1(L) \simeq \pi_1(X)$ we will have $\pi_2(f)$ f.g. over $Z\pi$.

2   HOMOTOPY TYPE OF FINITE-DIMENSIONAL (RESPECTIVELY FINITE) COMPLEXES

We will investigate conditions for X to be of the homotopy type of a CW-complex K with dim $K \leq n$. We will be concerned mainly with the case $n \geq 3$. If $X \sim K$ with dim $K \leq n$, then $H_i(\tilde{X}) = 0$ for $i > n$ and $H^{n+j}(X\colon \underline{B}) = 0$ for $j \geq 1$ and any local coefficient system $\underline{B}$ over X.

Definition 2.1

We say that X satisfies D_n if

    (i) $H_i(\tilde{X}) = 0$ for $i > n$.
    (ii) $H^{n+1}(X : \underline{B}) = 0$ for any local coefficient system $\underline{B}$ over X.

When X is homotopically equivalent to a complex K with dim $K \le n$, then we have already observed that X satisfies D_n. In this section we will show that if X satisfies D_n with $n \ge 3$, then X is homotopically equivalent to a complex of dimension less than or equal to n.

Proposition 2.2

Let $n \ge 3$ and $\varphi \colon K = K^{n-1} \to X$ be an $(n-1)$-connected map (K is not assumed to be finite). Suppose $H^{n+1}(X : \underline{B}) = 0$ for every local coefficient system $\underline{B}$ over X. Then $\pi_n(\varphi)$ is a projective $Z\pi$-module.

Proof. By Proposition 1.4, there exists a h.e. $\theta \colon \Gamma \to X$ extending φ with $\Gamma = K \cup$ (cells of dim $\ge n$). Up to homotopy $\varphi \colon K \to X$ could be replaced by the inclusion $K \to \Gamma$. Clearly $\Gamma^{n-1} = K$. Thus without loss of generality we can assume that φ equals the inclusion of X^{n-1} in X. Hence $\pi_n(\varphi) \simeq H_n(\tilde{\varphi}) = H_n(\tilde{X}, \tilde{X}^{n-1})$.

The cellular chain complex of the pair $(\tilde{X}, \tilde{X}^{n-1})$ is of the form

$$\cdots \to C_{n+2}(\tilde{X}, \tilde{X}^{n-1}) \to C_{n+1}(\tilde{X}, \tilde{X}^{n-1}) \to C_n(\tilde{X}, \tilde{X}^{n-1})$$

$$\to 0 \to 0 \to 0 \cdots .$$

Let $C_*(\tilde{X}) = \{C_k(\tilde{X}), d_k\}$ be the cellular chain complex of $\tilde{X}$. Then

$$C_i(\tilde{X}, \tilde{X}^{n-1}) = \begin{cases} C_i(\tilde{X}), & \text{for } i \ge n, \\ 0, & \text{for } i < n, \end{cases}$$

and the boundary map $C_{i+1}(\tilde{X}, \tilde{X}^{n-1}) \to C_i(\tilde{X}, \tilde{X}^{n-1})$ is the same as $d_{i+1} \colon C_{i+1}(\tilde{X}) \to C_i(\tilde{X})$ for $i \ge n$. It is clear that $C_{n+1}(\tilde{X}, \tilde{X}^{n-1})$ $\xrightarrow{d_{n+1}} C_n(\tilde{X}, \tilde{X}^{n-1}) \to H_n(\tilde{X}, \tilde{X}^{n-1}) \to 0$ is exact in $Z\pi$-mod.

Writing $B_n(\tilde{X})$ for Im $d_{n+1} \colon C_{n+1}(\tilde{X}) \to C_n(\tilde{X})$ and $j \colon B_n(\tilde{X}) \to C_n(\tilde{X})$ for the inclusion, we see that $0 \to B_n(\tilde{X}) \xrightarrow{j} C_n(\tilde{X}) \to \pi_n(\varphi) \to 0$ is exact in $Z\pi$-mod. Also $d_{n+1} \colon C_{n+1}(\tilde{X}) \to C_n(\tilde{X})$ could be factored as $d_{n+1} = j \circ c$ with $c \colon C_{n+1}(\tilde{X}) \to B_n(\tilde{X})$ an epimorphism.

The π-module $B_n(\tilde{X})$ yields a local coefficient system $\underline{B}$ over X. Then $c \in \text{Hom}_\pi(C_{n+1}(\tilde{X}), B_n(\tilde{X}))$ is an $(n+1)$ cochain of X with coefficients in

B. Moreover since $C_{n+1}(\tilde{X}) \xrightarrow{d_{n+2}} C_{n+1}(\tilde{X}) \xrightarrow{c} B_n(\tilde{X})$ has 0 as its composite map, we see that c is an $(n + 1)$ cocycle. Since $H^{n+1}(X; \underline{B}) = 0$, it follows that c is a coboundary. Hence $\exists$ an $s \in \mathrm{Hom}_\pi(C_n(\tilde{X}), B_n(\tilde{X}))$ with $c = sd_{n+1}$. Since $d_{n+1} = j \circ c$, we get $c = sj \circ c$. But $c: C_{n+1}(\tilde{X}) \to B_n(\tilde{X})$ is an epimorphism. Hence $sj = \mathrm{Id}_{B_n(\tilde{X})}$. Thus

$$0 \to B_n(\tilde{X}) \xrightarrow{j} C_n(\tilde{X}) \to \pi_n(\varphi) \to 0$$

is split by $s: C_n(\tilde{X}) \to B_n(\tilde{X})$. Hence $\pi_n(\varphi)$ is a direct summand of the free $Z\pi$-module $C_n(\tilde{X})$. $\square$

Proposition 2.3

Let X satisfy $H_1(\tilde{X}) = 0$ for $i > n$ where $n \geq 3$. Suppose $\varphi: K = K^{n-1} \to X$ is an $(n - 1)$-connected map with $\pi_n(\varphi)$ free over $Z\pi$. Let $\{\alpha_j\}_{j \in J}$ be a basis of $\pi_n(\varphi)$ over $Z\pi$ and

$$
\begin{array}{ccc}
S^{n-1} & \xrightarrow{f_j} & K \\
\downarrow & & \downarrow{\varphi} \\
D^n & \xrightarrow{g_j} & X
\end{array}
$$

represent α_j. Let $\Psi: L = K \cup_{f_j} \{e_j^n\}_{j \in J} \to X$ be the extension of φ described in Milnor's construction. Then Ψ is a h.e.

Proof. From Lemma 1.3 we know that Ψ is n-connected. Hence to show that Ψ is a h.e. we have only to show that $\pi_\mu(\Psi) = 0$ for $\mu > n$. By the Hurewicz theorem applied to $\tilde{\Psi}: \tilde{L} \to \tilde{X}$, this is equivalent to $H_\mu(\tilde{\Psi}) = 0$ for $\mu > n$. Thus we have only to show that $H_\mu(\tilde{\Psi}) = 0$ for $\mu > n$.

We have the exact homology sequence

$$\to H_i(\tilde{L}, \tilde{K}) \to H_i(\tilde{\varphi}) \to H_i(\tilde{\Psi}) \to H_{i-1}(\tilde{L}, \tilde{K}) \to \cdots.$$

From $H_i(\tilde{L}, \tilde{K}) = 0$ for $i \geq n + 1$ we see that $H_\mu(\tilde{\varphi}) \simeq H_\mu(\tilde{\Psi})$ for $\mu \geq n + 2$. Also in the commutative diagram

$$
\begin{array}{ccc}
\pi_n(L, K) \simeq \pi_n(\tilde{L}, \tilde{K}) & \xrightarrow[\simeq]{\rho} & H_n(\tilde{L}, \tilde{K}) \\
\hphantom{} & \text{Hurewicz} & \hphantom{} \\
\downarrow & \downarrow & \downarrow \\
\pi_n(\varphi) \quad \simeq \quad \pi_n(\tilde{\varphi}) & \xrightarrow[\text{Hurewicz}]{\rho_\simeq} & H_n(\tilde{\varphi})
\end{array}
$$

the map $\pi_n(L, K) \to \pi_n(\varphi)$ is an isomorphism. In fact $\pi_n(L, K)$ is free $Z\pi$ over $[\chi_j]$ as a basis, where χ_j is the characteristic map $(D^n, S^{n-1}) \to (L, K)$ for the cell e_j^n. We know that $[\chi_j]$ goes to $\alpha_j \in \pi_n(\varphi)$. By assumption α_j is a basis for $\pi_n(\varphi)$. Hence $\pi_n(L, K) \simeq \pi_n(\varphi)$. Hence, from the exactness of

$$H_{n+1}(\tilde{L}, \tilde{K}) \longrightarrow H_{n+1}(\tilde{\varphi}) \longrightarrow H_{n+1}(\tilde{\Psi}) \longrightarrow H_n(\tilde{L}, \tilde{K}) \overset{\simeq}{\longrightarrow} H_n(\tilde{\varphi})$$

with 0 mapping to $H_{n+1}(\tilde{L}, \tilde{K})$,

we see that $H_{n+1}(\tilde{\varphi}) \simeq H_{n+1}(\tilde{\Psi})$. In other words

$$\boxed{H_{n+i}(\tilde{\varphi}) \simeq H_{n+i}(\tilde{\Psi}) \quad \text{for all } i \geq 1.} \tag{1}$$

By assumption $H_\mu(\tilde{X}) = 0$ for $\mu > n$. Since $\dim K \leq n - 1$ we have $H_\nu(\tilde{K}) = 0$ for $\nu > n - 1$. The exact sequence

$$\cdots H_i(\tilde{K}) \overset{\tilde{\varphi}_*}{\longrightarrow} H_i(\tilde{X}) \longrightarrow H_i(\tilde{\varphi}) \longrightarrow H_{i-1}(\tilde{K}) \longrightarrow \cdots$$

now shows that $H_\mu(\tilde{\varphi}) = 0$ for $\mu > n$. From (1) we get $H_\mu(\tilde{\Psi}) = 0$ for $\mu > n$. This proves that $\Psi: L \to X$ is a h.e. $\square$

Lemma 2.4

Let P be any projective module over a ring R. Then there exists a free R-module F with $P \oplus F \simeq F$.

Proof. Since P is projective, there exists an R-module Q with $P \oplus Q$ free, say F'. Let $F_k' = F'$ for each $k \geq 1$ and $F = \oplus_{k \geq 1} F_k'$. Then defining $P_k = P$, $Q_k = Q$, for $k \geq 1$, we have $F = \oplus_{k \geq 1}(P_k \oplus Q_k) \simeq P_1 \oplus (\oplus_{k \geq 1} C_k)$ where

$$C_k = Q_k \oplus P_{k+1} = Q \oplus P \simeq F' = F_k'.$$

Thus $F \simeq P \oplus (\oplus_{k \geq 1} F_k') = P \oplus F$. $\square$

Theorem 2.5

Let $n \geq 3$. Then X is homotopically equivalent to a complex L with $\dim L \leq n \Leftrightarrow X$ satisfies D_n.

Proof. If $X \sim L$ with $\dim L \leq n$, then $H_i(\tilde{L}) = 0$ for $i > n$ and $H^{n+j}(X, \underline{B}) = 0$ for any local system $\underline{B}$ over X and any $j \geq 1$. In particular X satisfies D_n.

Assume X satisfies D_n. By Remark 1.13 there exists an $(n - 1)$-connected map $\varphi: K = K^{n-1} \to X$. From Proposition 2.2 we see that $\pi_n(\varphi)$ is a projective module over $Z\pi$. From Lemma 2.4 we can get a free $Z\pi$-module F with $\pi_n(\varphi) \oplus F \simeq F$. Let $\{x_j\}_{j \in J}$ denote a basis of F over $Z\pi$. Let $K' = K \vee (\vee_{j \in J} S_j^{n-1})$ and $\varphi': K' \to X$ be defined by $\varphi'|K = \varphi$, $\varphi'(S_j^{n-1}) = * \in X$. If $\mu: K \to X'$ denotes the inclusion, since $n \geq 3$ we see that $\mu_*: \pi_1(K)' \simeq \pi_1(K')$. Hence $\varphi'_*: \pi_1(K') \simeq \pi_1(X)$. If $\tilde{\varphi}': \tilde{K}' \to \tilde{X}$ denotes a lift of φ' to universal coverings, then $\tilde{K} = p^{-1}(K)$ is the universal covering of K where $p: \tilde{K}' \to K'$ is the covering projection and $\tilde{\varphi}'|\tilde{K} = \tilde{\varphi}$ is a lift of φ. Let $\tilde{\mu}: \tilde{K} \to \tilde{K}'$ denote the inclusion. Then we have an exact sequence

$$\to H_k(\tilde{K}'\tilde{K}) \to H_k(\tilde{\varphi}) \to H_k(\tilde{\varphi}') \to H_{k-1}(\tilde{K}', \tilde{K}) \to \cdots . \quad (**)$$

We know that $H_k(\tilde{K}', \tilde{K}) = 0$ for $k \neq n - 1$ and $H_{n-1}(\tilde{K}', \tilde{K})$ is free $Z\pi$ on a basis indexed by $J \simeq F$. We know that $H_k(\tilde{\varphi}) = 0$ for $k \leq n - 1$. It follows from $(**)$ that $H_k(\tilde{\varphi}') = 0$ for $k \leq n - 1$. Also

$$H_n(\tilde{K}', \tilde{K}) \longrightarrow H_n(\tilde{\varphi}) \longrightarrow H_n(\tilde{\varphi}') \longrightarrow H_{n-1}(\tilde{K}', \tilde{K}) \longrightarrow H_{n-1}(\tilde{\varphi})$$

$$\|$$
$$0 \qquad\qquad\qquad\qquad\qquad\qquad\qquad\qquad \downarrow \simeq \qquad\qquad \|$$
$$\qquad\qquad\qquad\qquad\qquad\qquad\qquad\qquad F \qquad\qquad 0$$

is exact. Since F is free $Z\pi$, we get $H_n(\tilde{\varphi}') \simeq H_n(\tilde{\varphi}) \oplus F$. Since $H_k(\tilde{\varphi}') = 0$ for $k \leq n - 1$, we get

$$\pi_n(\tilde{\varphi}') \simeq H_n(\tilde{\varphi}')$$

$$\simeq H_n(\tilde{\varphi}) \oplus F$$

$$\simeq \pi_n(\varphi) \oplus F$$

$$\simeq F.$$

Thus $\varphi': K' \to X$ is an $(n-1)$-connected map with $\pi_n(\varphi')$ free as a $Z\pi$-module. From Proposition 2.3 it now follows that φ' can be extended to a h.e. $\Psi: K' \cup \{e_j^n\}_{j \in J} \to X$. This completes the proof of Theorem 2.5. $\square$

Corollary 2.6

Let $n \geq 3$ and $1 \leq r \leq n - 2$. Suppose X satisfies D_n and F_r. Then there exists a h.e. $\theta: L \to X$ with $\dim L \leq n$ and L^r finite.

Proof. Since X satisfies F_r, $\exists$ a h.e. $\alpha: A \to X$ with A a CW-complex having a finite r-skeleton. If $K = A^{n-1}$ and $\varphi = \alpha|A^{n-1}$, then $\varphi: K = K^{n-1} \to$

X is an $(n - 1)$-connected map. [In fact up to homotopy φ could be regarded as the inclusion $A^{n-1} \to A$ and is hence $(n - 1)$-connected.]

The proof of Theorem 2.5 yields a h.e. $\Psi: L = (K \vee (\vee_{j \in J} S_j^{n-1})) \cup \{e_j^n\}_{j \in J} \to X$. This complex L satisfies $\dim L \le n$ and L^r finite.

Let $n \ge 3$ and suppose X satisfies D_n and F_n. Since X satisfies F_{n-1} there exists an $(n - 1)$-connected map $K = K^{n-1} \overset{\varphi}{\to} X$ with K finite. Since X satisfies D_n we see that $\pi_n(\varphi)$ is a projective $Z\pi$-module. Since X satisfies F_n, $\pi_n(\varphi)$ is f.g. over $Z\pi$. Let $\Lambda = Z\pi$. Then $P = \pi_n(\varphi)$ determines an element $[P] \in \tilde{K}_0(\Lambda)$.

We can find an epimorphism $\varepsilon: F \to P$ of a free Λ-module with a finite basis, say $\{x_i\}_{i=1}^k$. Then if $\alpha_i = \varepsilon(x_i)$, P is generated by α_i. If

$$\begin{array}{ccc} S^{n-1} & \overset{f_i}{\longrightarrow} & K \\ \downarrow & & \downarrow \varphi \\ D^n & \overset{g_i}{\longrightarrow} & X \end{array}$$

represents $\alpha_i \in \pi_n(\varphi)$, then the extension $\Psi: L = K^{n-1} \cup_{f_i} \{e_j^n\}_{i=1}^k \to X$ of φ given by Milnor's construction is an n-connected map and $L = L^n$ is finite. From the commutative diagram

$$\begin{array}{ccc} \pi_n(L, K) & \longrightarrow & \pi_n(\varphi) \\ \downarrow {\scriptstyle \simeq} & & \downarrow \\ H_n(\tilde{L}, \tilde{K}) & \longrightarrow & H_n(\tilde{\varphi}) \end{array}$$

we see that $\pi_n(L, K)$ is free $Z\pi$ on $[\chi_i]$ as basis, where $\chi_i: (D^n, S^{n-1}) \to (L, K)$ is the characteristic map for the n cell e_i^n now that χ_i gets carried to $\alpha_i \in \pi_n(\varphi)$. Thus we can identify $H_n(\tilde{L}, \tilde{K}) \to H_n(\tilde{\varphi})$ with the map $F \overset{\varepsilon}{\to} P$. Since X satisfies D_n, we have $H_i(\tilde{X}) = 0$ for $i > n$. Also $\dim K \le n - 1 \Rightarrow H_i(\tilde{K}) = 0$ for $i > n - 1$. The exactness of $\cdots \to H_i(\tilde{K}) \to H_i(\tilde{X}) \to H_i(\tilde{\varphi}) \to H_{i-1}(\tilde{K}) \to \cdots$ shows that $H_i(\tilde{\varphi}) = 0$ for $i > n$. From the exact sequence

$$\begin{array}{ccccccccc} H_{n+1}(\tilde{\varphi}) & \longrightarrow & H_{n+1}(\tilde{\Psi}) & \longrightarrow & H_n(\tilde{L}, \tilde{K}) & \longrightarrow & H_n(\tilde{\varphi}) & \longrightarrow & H_n(\tilde{\Psi}) \\ \| & & & & \| & & \| & & \| \\ 0 & & & & F & \overset{\varepsilon}{\longrightarrow} & P & & 0 \end{array}$$

and the projective nature of P we see that $H_{n+1}(\tilde{\Psi}) \oplus P \simeq F$. Let $H_{n+1}(\tilde{\Psi}) = Q$. Then $Q \oplus P \simeq F$.

It follows that $[\pi_{n+1}(\tilde{\Psi})] = [\pi_{n+1}(\Psi)] = [Q] = -[P] \in \tilde{K}_0(\Lambda)$. $\square$

In what follows n denotes an integer greater than or equal to 3.

Lemma 2.7

Let X satisfy D_n and $\Psi: L \to X$ be n-connected. Then $\exists$ an $s: X \to L$ with $\Psi \circ s \sim \mathrm{Id}_X$. Hence L dominates X.

Proof. Replace Ψ by a fibration. If F denotes the fiber of Ψ we have $\pi_{i-1}(F) \simeq \pi_i(\Psi)$. The obstructions to finding a cross section $s: X \to L$ to $\Psi: L \to X$ are in $H^i(X, \pi_{i-1}(F)) \simeq H^i(X; \pi_i(\Psi))$. For $i \leq n$ we have $\pi_i(\Psi) = 0$. Also for $j \geq 1$ we have $H^{n+j}(X, \pi_{n+j}(\Psi)) = 0$ since X is homotopically equivalent to an n-dimensional complex by Theorem 2.5.

Hence a cross section $s: X \to L$ exists. $\square$

Remark 2.8

Suppose X satisfies F_n and D_n. Then the map $\Psi: L^n \to X$ described before Lemma 2.7 satisfies the conditions: (i) L is finite and (ii) Ψ is n-connected and $\pi_{n-1}(\Psi)$ is f.g. projective over $Z\pi$. From Remark 1.16 it follows that X satisfies F_{n+1}. Also from Theorem 2.5 we see that X is homotopically equivalent to a complex of dimension less than or equal to n. Hence X satisfies D_{n+1} as well.

Lemma 2.9

Let X satisfy F_n and D_n. Let $L_i = L_i^n$ be finite, $\Psi_i: L_i \to X$ be n-connected, and $Q_i = \pi_{n+1}(\Psi_i)$, $i = 1, 2$. Then $[Q_1] = [Q_2]$ in $\tilde{K}_0(\Lambda)$. (Observe that X also satisfies F_{n+1} and D_{n+1} and hence Q_1, Q_2 are f.g. projective over $Z\pi$.)

Proof. From Lemma 2.7 $\exists$ $s_i: X \to L_i$ with $\Psi_i s_i \sim \mathrm{Id}_X$. Hence $H_j(\tilde{X}) \xrightarrow{\tilde{s}_{i*}} H_j(\tilde{L}_i) \xrightarrow{\tilde{\Psi}_{i*}} H_j(\tilde{X})$ has $\mathrm{Id}_{H_j(\tilde{X})}$ as its composite. Since X satisfies D_n, $H_j(\tilde{X}) = 0$ for $j > n$. Since dim $L_i \leq n$, we have $H_j(\tilde{L}_i) = 0$ for $j > n$. Since Ψ_i is n-connected, we get $H_j(\tilde{\Psi}_i) = 0$ for $j < n$ and $Q_i = \pi_{n+1}(\Psi_i) \simeq H_{n+1}(\tilde{\Psi}_i)$. From the exact sequence

$$\longrightarrow H_j(\tilde{L}_i) \xrightarrow{\tilde{\Psi}_{i*}} H_j(\tilde{X}) \longrightarrow H_j(\tilde{\Psi}_i) \longrightarrow H_{j-1}(\tilde{L}_i) \longrightarrow$$

we now get

$$H_j(\tilde{L}_i) \xrightarrow[\simeq]{\tilde{\Psi}_{i*}} H_j(\tilde{X}) \quad \text{for } j < n$$

and

$$0 \longrightarrow H_{n+1}(\tilde{\Psi}_i) \longrightarrow H_n(\tilde{L}_i) \xrightarrow{\tilde{\Psi}_{i*}} H_n(\tilde{X}) \longrightarrow 0 \quad \text{split exact,}$$

split by $H_n(\tilde{L}_i) \xrightarrow{\tilde{s}_{i*}} H_n(\tilde{X})$. Thus

$$H_n(\tilde{L}_i) \simeq H_n(\tilde{X}) \oplus Q_i, \qquad i = 1, 2.$$

Let $h: L_1 \to L_2$ be the map $h = s_2 \Psi_1$. When we identify $H_n(\tilde{L}_1)$ with $H_n(\tilde{X}) \oplus Q_1$, $\tilde{\Psi}_{1*}$ corresponds to the projection of $H_n(\tilde{X}) \oplus Q_1$ onto $H_n(\tilde{X})$. When we identify $H_n(\tilde{L}_2)$ with $H_n(\tilde{X}) \oplus Q_2$ the map $\tilde{s}_{2*}: H_n(\tilde{X}) \to H_n(\tilde{X}) \oplus Q_2$ corresponds to the obvious inclusion. Thus $\tilde{h}_*: H_n(\tilde{L}_1) \to H_n(\tilde{L}_2)$ corresponds to the map $\alpha: H_n(\tilde{X}) \oplus Q_1 \to H_n(\tilde{X}) \oplus Q_2$ given by $\alpha(x, c) = (x, 0)$ for any $x \in H_n(\tilde{X})$ and $c \in Q_1$.

Observe that $h: L_1 \to L_2$ induces an isomorphism at π_1 level. Up to homotopy h could be replaced by the inclusion $\mu: L_1 \to M_h$ of L_1 into the mapping cylinder of h. We could have chosen X to be a CW-complex, and s_i, Ψ_i cellular and hence h cellular. Thus M_h is a CW-complex of dim $\leq n + 1$ (since dim $L_i \leq n$). Also M_h is finite.

From $H_j(\tilde{L}_i) \xrightarrow[\simeq]{\tilde{\Psi}_{i*}} H_j(\tilde{X})$ for $j < n$ and $\tilde{s}_{i*}$ an inverse of $\tilde{j}_{i*}$ we see that $\tilde{s}_{i*}: H_j(\tilde{X}) \simeq H_j(\tilde{L}_i)$ for $j < n$. Hence $\tilde{h}_*: H_j(\tilde{L}_1) \simeq H_j(\tilde{L}_2)$ for $j < n$. Hence $H_j(\tilde{h}_*) = 0$ for $j < n$. Also dim $L_i \leq n \Rightarrow H_j(\tilde{L}_i) = 0$ for $j > n$. These in turn yield the exact sequence

$$H_{n+1}(\tilde{L}_2) \cdots \longrightarrow H_{n+1}(\tilde{h}) \longrightarrow H_n(\tilde{L}_1) \xrightarrow{\tilde{h}_*} H_n(\tilde{L}_2) \longrightarrow H_n(\tilde{h}) \longrightarrow \cdots$$
$$\parallel$$
$$0$$

$$\cdots \longrightarrow H_{n-1}(\tilde{L}_1) \xrightarrow[\simeq]{\tilde{h}_*} H_{n-1}(\tilde{L}_2)$$

or

$$0 \longrightarrow H_{n+1}(\tilde{h}) \longrightarrow H_n(\tilde{L}_1) \xrightarrow{\tilde{h}_*} H_n(\tilde{L}_2) \longrightarrow H_n(\tilde{h}) \longrightarrow 0$$

exact. But $H_n(\tilde{L}_1) \xrightarrow{\tilde{h}_*} H_n(\tilde{L}_2)$ could be identified with $\alpha: H_n(\tilde{X}) \oplus Q_1 \to H_n(\tilde{X}) \oplus Q_2$ where $\alpha(x, c) = (x, 0)$. Hence $H_{n+1}(\tilde{h}) \simeq Q_1$ and $H_n(\tilde{h}) \simeq Q_2$. Replacing h by $\mu: L_1 \to M_h$ we get $H_j(\tilde{\mu}) = 0$ for $j < n$, $H_n(\tilde{\mu}) \simeq Q_2$ and $H_{n+1}(\tilde{\mu}) \simeq Q_1$. The cellular chain complex of $(\tilde{M}_h, \tilde{L}_1)$ is of the form $0 \to C_{n+1}(\tilde{M}_h, \tilde{L}_1) \to \cdots \to C_0(\tilde{M}_h, \tilde{L}_1) \to 0$. If $Z_k(\tilde{M}_h, \tilde{L}_1) = \ker d_k: C_k(\tilde{M}_h, \tilde{L}_1) \to C_k(\tilde{M}_h, \tilde{L}_1)$ and $b_k(\tilde{M}_h, \tilde{L}_1) = \operatorname{Im} d_{k+1}: C_{k+1}(\tilde{M}_h, \tilde{L}_1) \to C_k(\tilde{M}_h, \tilde{L}_1)$, we have $Z_k(\tilde{M}_h, \tilde{L}_1) = B_k(\tilde{M}_h, \tilde{L}_1)$ for $k < n$. Also we have the following exact sequences

$$0 \to Z_1 \to C_1 \to C_0 \to 0, \qquad\qquad (1)$$

$$0 \to Z_2 \to C_2 \to Z_1 \to 0, \qquad\qquad (2)$$

$$\cdots$$

$$0 \to Z_n \to C_n \to Z_{n-1} \to 0, \qquad\qquad (n-1)$$

$$0 \to B_n \to Z_n \to Q_2 \to 0, \qquad\qquad (n)$$

$$0 \to Q_1 \to C_{n+1} \to B_n \to 0. \qquad\qquad (n+1)$$

Using the fact that C_0 is free we see that (1) splits and hence Z_1 is f.g. projective over $Z\pi$. Now (2) splits. Proceeding thus we see that the sequences $(1), \ldots, (n-1)$ split and that Z_k is f.g. projective over $Z\pi$ for $k \leq n$.

The sequence (n) splits since Q_2 is projective. It follows that B_n is projective and hence $(n+1)$ splits.

Using successively the exact sequences (1) up to $(n-1)$ we get $[Z_1] = 0$ and $[Z_1] + [Z_2] = 0, \ldots, [Z_{n-1}] + [Z_n] = 0$ in $\tilde{K}_0(\Lambda)$. Hence $[Z_n] = 0$ in $\tilde{K}_0(\Lambda)$. The sequence (n) yields $[B_n] + [Q_2] = [Z_n] = 0$ in $\tilde{K}_0(\Lambda)$. Sequence $(n+1)$ yields $[Q_1] + [B_n] = 0$ in $\tilde{K}_0(\Lambda)$. Hence $[Q_1] = -[B_n] = [Q_2]$. This completes the proof of Lemma 2.9. $\square$

Theorem 2.10

X is dominated by a finite complex $\Leftrightarrow$ X satisfies F_{n_0} and D_{n_0} for some $n_0 \geq 3$. In this case X satisfies F_n and D_n for all $n \geq n_0$. Moreover given any $n \geq n_0$ we can find an n-connected map $\Psi: L = L^n \to X$ with L finite. In this case $\pi_{n+1}(\Psi)$ is f.g. projective over $Z\pi$ and the element $\tilde{w}(X) = (-1)^{n+1}[\pi_{n+1}(\Psi)]$ depends only on the homotopy type of X. The vanishing of $\tilde{w}(X)$ is necessary and coefficient for X to be homotopy equivalent to a finite complex.

Proof. Let X be dominated by a finite complex $K = K^k$. Then for any $n_0 \geq k$ we have $K = K^{n_0}$ is finite. From (ii) $\Rightarrow$ (iii) of Theorem 1.11 we see that X satisfies F_{n_0}. It is clear that X satisfies D_{n_0} for any $n_0 \geq k$.

Conversely, suppose X satisfies F_{n_0} and D_{n_0} with $n_0 \geq 3$. Then $\exists$ an n_0-connected map $K = K^{n_0} \overset{\alpha}{\to} X$ with K finite. From Lemma 2.7 we see that K dominates X. Also the comments preceding Lemma 2.7 together with Lemma 2.9 show that $(-1)^{n+1}[\pi_{n+1}(\Psi)]$ depends only on the homotopy type of X.

If X is homotopically equivalent to a finite complex, then we can assume that X itself is a finite complex. Since $X \sim X \vee D^n$, we can even assume that $\dim X = n \geq 3$. Thus for $\Psi: L = L^n \to X$ we can choose Id_X. Hence $\pi_{n+1}(\Psi) = 0$ or $\tilde{w}(X) = 0$.

Conversely, assume that $\tilde{w}(X) = 0$. Then $\exists$ a f.g. free $Z\pi$-module F such that $\pi_{n+1}(\Psi) \oplus F$ is free. Let $\{a_j\}_{j=1}^r$ be a basis for F and $\theta: L \vee \{\vee S_j^n\}_{j=1}^r$

$\rightarrow X$ be given by $\theta|L = \Psi$, $\theta(S_j^n) = *$. As seen in the proof of Theorem 2.5, we have θ to be n-connected and $\pi_{n+1}(\theta) \simeq \pi_{n+1}(\Psi) \oplus F$. Hence $\pi_{n+1}(\theta)$ is f.g.-free over $Z\pi$. If $\{x_i\}_{i=1}^\ell$ is a basis for $\pi_{n+1}(\theta)$, then Proposition 2.3 yields a homotopy equivalence $f: (L \vee (\vee_{j=1}^r S_j^n)) \cup \{e_i^{n+1}\}_{i=1}^\ell \rightarrow X$ extending θ. This completes the proof of Theorem 2.10. $\square$

Theorem 2.11

Let $n \geq 3$ and $K = K^{n-1}$ be any finite complex of dimension less than or equal to $n - 1$, $\pi = \pi_1(K)$. Let $x \in \tilde{K}_0(Z\pi)$ be any given element. Then $\exists$ a finitely dominated CW-complex X with $X^{n-1} = K$ and $\tilde{w}(X) = x$.

Proof. Let A be a f.g. projective $Z\pi$-module with $(-1)^n x = [A]$. $\exists$ a f.g. projective $Z\pi$-module B with $A \oplus B = F$ free over $Z\pi$. Let $p_A: F \rightarrow A$ and $p_B: F \rightarrow B$ denote the respective projections. We choose a fixed basis $\{\alpha_i\}_{i=r}^r$ for F over $Z\pi$.

Let $L = L^n = K \vee (\vee_{i=1}^r S_i^n)$. Let e denote the identity element of π and $\tilde{S}_{i,e}^n$ be the lift of S_i^n with the base point of S_i^n lifted to a given fixed lift of the base point of L to $\tilde{L}$. Let $\tilde{S}_{i,\sigma}^n$ denote the translate of $\tilde{S}_{i,e}^n$ by the action of the element $\sigma \in \pi$ on $\tilde{L}$. Then $\tilde{L}$ is homotopically equivalent to $\tilde{K} \vee (\vee_{i=1, \sigma \in \pi}^r \tilde{S}_{i,\sigma}^n)$.

A, B are projective over $Z\pi$ and hence free over Z. Let $\{a_\mu\}_{\mu \in I}$ and $\{b_\nu\}_{\nu \in J}$ be bases for A, B over Z chosen once and for all. We have

$$\pi_n(L) \simeq \pi_n(\tilde{L}_n) \simeq \pi_n(\tilde{K}) \oplus \pi_n(\tilde{L}, \tilde{K})$$

$$\simeq \pi_n(K) \oplus F \quad (\text{in } Z\pi\text{-mod})$$

$$\simeq \pi_n(K) \oplus A \oplus B.$$

Thus a_μ (resp. b_ν) could be regarded as an element of A (resp. B) with $A \subset \pi_n(\tilde{L})$ [resp. $B \subset \pi_n(\tilde{L})$]. Let $f_\mu: S^n \rightarrow \tilde{L}$ represent a_μ, b_ν, respectively. Let $h: \tilde{K} \vee (\vee_{\mu \in I} S_\mu^n) \vee (\vee_{\nu \in J} S_\nu^n) \rightarrow \tilde{L}$ be defined by $h|\tilde{K}$ equal to the inclusion of $\tilde{K}$ in $\tilde{L}$, $h|S_\mu^n = f_\mu$, and $h|S_\nu^n = g_\nu$. From $H_n(\tilde{L}, \tilde{K}) = F$ with $\tilde{S}_{i,e}^n$ representing $\alpha_i \in F = H_n(\tilde{L}, \tilde{K})$ we see immediately that $h: \tilde{K} \vee (\vee_{\mu \in I} S_\mu^n) \vee (\vee_{\nu \in J} S_\nu^n) \rightarrow \tilde{L}$ induces isomorphism in homology. Since $\tilde{K} \vee (\vee_{\mu \in I} S_\mu^n) \vee (\vee_{\nu \in J} S_\nu^n)$ and $\tilde{L}$ are both simply connected it follows that h is a homotopy equivalence. A word of caution at this time may not be out of place. While π acts on $\tilde{L}$ as the Deck transformation group, there is no action of π on $\tilde{K} \vee (\vee_{\mu \in I} S_\mu^n) \vee (\vee_{\nu \in J} S_\nu^n)$. However π acts on $\tilde{K}$ and $h|\tilde{K}$ is equal to the inclusion of $\tilde{K}$ in $\tilde{L}$ and hence $h|\tilde{K}$ preserves the action of π.

The cellular chain complex of $\tilde{L}$ is of the form

$$\cdots \rightarrow 0 \rightarrow 0 \rightarrow C_n(\tilde{L}) \rightarrow C_{n-1}(\tilde{L}) \rightarrow \cdots \rightarrow C_0(\tilde{L}) \rightarrow 0 \rightarrow 0 \rightarrow \cdots$$

with $C_n(\tilde{L}) = F$ and $C_i(\tilde{L}) = C_i(\tilde{K})$ for $i \leq n - 1$. The cellular chain complex of $\tilde{K} \vee (\vee S_\mu^n) \vee (\vee_\nu S_\nu^n)$ is of the form

$$\cdots 0 \to 0 \to A \oplus B \to C_{n-1}(\tilde{K}) \to \cdots \to \left(C_0(\tilde{K}) \right) \to 0 \to 0 \to \cdots.$$

The map induced by h is an isomorphism

$$C_*(\tilde{K} \vee (\vee_\mu S_\mu^n) \vee (\vee_\nu S_\nu^n)) \cdots 0 \to A \oplus B \to C_{n-1}(\tilde{K}) \to \cdots C_0(\tilde{K}) \to 0 \to \cdots$$
$$\downarrow \simeq$$
$$C_*(\tilde{L}): \qquad\qquad \cdots 0 \to F \to C_{n-1}(\tilde{K}) \to \cdots \to C_0(\tilde{K}) \to 0 \to 0 \to .$$

which is actually the identity map.

Let $W = \tilde{K} \vee (\vee_{\mu \in I} S_\mu^n)$ and $u = h|W: W \to \tilde{L}$. Then u induces the inclusion of the cellular chain complexes

$$C_*(W): \cdots \to 0 \to A \to C_{n-1}(\tilde{K}) \to \cdots \to C_0(\tilde{K}) \to 0 \to 0 \to \cdots$$
$$\downarrow \mu_* \qquad\qquad\qquad \downarrow \qquad\qquad \| \qquad\qquad\qquad \|$$
$$C_*(\tilde{L}): \cdots \to 0 \to F \to C_{n-1}(\tilde{K}) \to \cdots \to C_0(\tilde{K}) \to 0 \to 0 \to \cdots.$$

Though there is no action of π on W, the cellular chain complex $C_*(W)$ is in a natural way a chain complex over $Z\pi$ and $u_*: C_*(W) \to C_*(\tilde{L})$ is a $Z\pi$-chain map. This fact will play a crucial role.

We will now construct a CW-complex X (infinite dimensional but having finitely many cells in each dimension) which will satisfy the following requirements:

(i) $X^n = L$.

(ii) $\exists$ a homotopy equivalence $h_k: W \vee (\vee_{\nu \in J} S_\nu^{n+2k}) \to \tilde{X}^{n+2k}$ (for any integer $k \geq 0$) with $h_k|W = j_k \circ u$ where $j_k: \tilde{X}^n \to \tilde{X}^{n+2k}$ is the inclusion and $h_{k*}: \pi_{n+2k}(W) \oplus B \to \pi_{n+2k}(\tilde{X}^{n+2k})$ is a $Z\pi$ isomorphism.

In fact by induction on k, we will construct the $n + 2k$ skeleton of X so as to satisfy (i) and (ii). For $k = 0$, we know that $h_0 = h: W \vee (\vee_{\nu \in J} S_\nu^n) \to \tilde{X}^n = \tilde{L}$ is a homotopy equivalence.

Let $k \geq 0$ and assume X^{n+2k} constructed so as to satisfy (i) and (ii). Let $c_i = p_B(\alpha_i) \in B$. From condition (ii) we see that $\pi_{n+2k}(\tilde{X}^{n+2k}) \simeq \pi_{n+2k}(W) \oplus B$. Thus $\pi_{n+2k}(X^{n+2k}) \simeq \pi_{n+2k}(W) \oplus B$. Let $\theta_i: S^{n+2k} \to X^{n+2k}$ represent $c_i \in B \subset \pi_{n+2k}(X^{n+2k})$ and let $X^{n+2k+1} = X^{n+2k} \cup_{\theta_i} \{e_i^{n+2k+1}\}_{i=1}^r$.

Consider the homotopy, homology ladder in Diagram C. The maps ρ are all Hurewicz homomorphisms. By construction, the map

$$\partial: \pi_{n+2k+1}(\tilde{X}^{n+2k+1}, \tilde{X}^{n+2k}) \to \pi_{n+2k}(\tilde{X}^{n+2k}) = \pi_{n+2k}(W) \oplus B$$

is the same as the map $\alpha \mapsto (0, p_B(\alpha))$ of F in $\pi_{n+k}(W) \oplus B$. It follows that the map $\partial: H_{n+2k+1}(\tilde{X}^{n+2k+1}, \tilde{X}^{n+2k}) \to H_{n+2k}(\tilde{X}^{n+2k}) = H_{n+2k}(W) \oplus B$ can be identified with the map $\alpha \mapsto (0, p_B(\alpha))$ of F in $H_{n+2k}(W) \oplus B$.

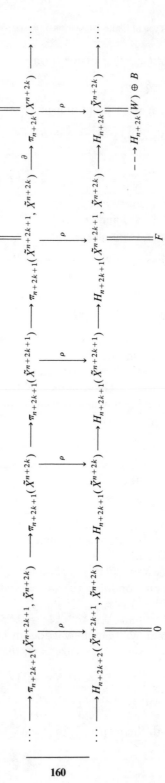

Diagram C.

Writing j_* for both the maps

$$\pi_{n+2k+1}(\tilde{X}^{n+2k+1}) \to \pi_{n+2k+1}(\tilde{X}^{n+2k+1}, \tilde{X}^{n+2k}),$$

$$\left(H_{n+2k+1}(\tilde{X}^{n+2k+1}) \to H_{n+2k+1}(\tilde{X}^{n+2k+1}, \tilde{X}^{n+2k})\right)$$

from Diagram C, we now get the commutative diagram of exact sequences in $Z\pi$-mod.

$$\longrightarrow \pi_{n+2k+1}(\tilde{X}^{n+2k}) \longrightarrow \pi_{n+2k+1}(\tilde{X}^{n+2k+1}) \xrightarrow{j_*} A \longrightarrow 0$$

$$\Big\downarrow \rho \qquad\qquad \Big\downarrow \rho \qquad\qquad \Big\|$$

$$0 \longrightarrow H_{n+2k+1}(\tilde{X}^{n+2k}) \longrightarrow H_{n+2k+1}(\tilde{X}^{n+2k+1}) \xrightarrow{j_*} A \longrightarrow 0$$

Diagram D.

Since A is $Z\pi$-projective $\exists$ a splitting

$$t: A \to \pi_{n+2k+1}(\tilde{X}^{n+2k+1}) \quad \text{for } j_*: \pi_{n+2k+1}(\tilde{X}^{n+2k+1}) \twoheadrightarrow A.$$

If $s = \rho t: A \to H_{n+2k+1}(\tilde{X}^{n+2k+1})$, then s splits $j_*: H_{n+2k+1}(\tilde{X}^{n+2k+1}) \to A$. Using the splitting s we can identify $H_{n+2k+1}(\tilde{X}^{n+2k+1})$ with $H_{n+2k+1}(\tilde{X}^{n+2k}) \oplus A$. Moreover under this identification $A \subset$ Im $\rho: \pi_{n+2k+1}(\tilde{X}^{n+2k+1}) \to H_{N+2k+1}(\tilde{X}^{n+2k})$. In other words all the elements in $A \subset H_{n+2k+1}(\tilde{X}^{n+2k+1})$ are spherical.

The exactness of

$$H_{n+2k+1}(\tilde{X}^{n+2k+1}, \tilde{X}^{n+2k}) \longrightarrow H_{n+2k}(\tilde{X}^{n+2k}) \longrightarrow H_{n+2k}(\tilde{X}^{n+2k+1}) \longrightarrow H_{n+2k}(\tilde{X}^{n+2k+1}, \tilde{X}^{n+2k}) \longrightarrow$$

$$\Big\| \qquad\qquad\qquad \Big\| \qquad\qquad\qquad \Big\|$$

$$F \dashrightarrow H_{n+2k}(W) \oplus B \qquad\qquad\qquad\qquad 0$$

$$\alpha \longrightarrow (0, p_B(\alpha))$$

shows

$$H_{n+2k}(\tilde{X}^{n+2k+1}) \simeq H_{n+2k}(W),$$

$$H_i(\tilde{X}^{n+2k+1}) \simeq H_i(\tilde{X}^{n+2k}) \simeq H_i(W) \quad \text{for } i < n + 2k.$$

Let $f_\mu^{(k)}: S^{n+2k+1} \to \tilde{X}^{n+2k+1}$ represent $a_\mu \in A \subset H_{n+2k+1}(\tilde{X}^{n+2k+1})$. If $\gamma: W \vee (\vee_{\mu \in I} S_\mu^{n+2k+1}) \to \tilde{X}^{n+2k+1}$ is given by $\gamma|W = i \circ u$ where $i: \tilde{X}^{n+2k} \hookrightarrow$

$\tilde{X}^{n+2k+1}$ and $\gamma|S_\mu^{n+2k+1} = f_\mu^{(k)}$, then γ induces isomorphisms in homology. Since the spaces involved are simply connected we see that γ is a h.e. Using γ_* we see that $\pi_{n+2k+1}(\tilde{X}^{n+2k+1}) \simeq \pi_{n+2k+1}(W) \oplus A$. If $d_i = p_A(\alpha_i) \in A \subset \pi_{n+2k+1}(\tilde{X}^{n+2k+1})$ and if $\varphi_i: S^{n+2k+1} \to X^{n+2k+1}$ represents d_i, then arguing as before we get a homotopy equivalence $h_{k+1}: W \vee (\vee_{\nu \in J} S_\nu^{n+2k+2}) \to \tilde{X}^{n+2k+2}$ with $h_{k+1}|W = u$ where $X^{n+2k+2} = X^{n+2k+1} \cup_{\varphi_i} \{e_i^{n+2k+2}\}_{i=1}^r$.

This completes the inductive step in the construction of the complex X. From (ii) we see that

$$u_*: \pi_i(W) \to \pi_i(\tilde{X}) \quad \text{is an isomorphism for all } i.$$

In fact

$$u_*: \pi_i(W) \to \pi_i(\tilde{X}^{n+2k}) \quad \text{is an isomorphism for } i \le n + 2k - 1$$

from (ii). But (ii) is valid for all k and

$$\pi_i(\tilde{X}^{n+2k}) \simeq \pi_i(\tilde{X}) \quad \text{for } i \le n + 2k - 1.$$

It follows that $u: W \to \tilde{X}$ is a homotopy equivalence or $u: \tilde{K} \vee (\vee_{\mu \in I} S_\mu^n) \to \tilde{X}$ is a homotopy equivalence. Since $X^n = L$ we see that the cellular chain complex $C_*(\tilde{X})$ is of the form

$$\cdots \to C_{n+2}(\tilde{X}) \to C_{n+1}(\tilde{X}) \to F \to C_{n-1}(\tilde{K}) \to \cdots \to C_0(\tilde{K}) \to 0 \to 0 \to$$

and the map $u_*: C_*(W) \to C_*(\tilde{X})$ is described next:

$$C_*(W): \to \quad 0 \quad \to A \to C_{n-1}(\tilde{K}) \to \cdots \to C_0(\tilde{K}) \to 0 \to 0 \to$$
$$\downarrow u_* \qquad \downarrow \qquad \qquad \| \qquad \qquad \qquad \|$$
$$C_*(\tilde{X}): \to C_{n+2}(\tilde{X}) \to F \to C_{n-1}(\tilde{K}) \to \cdots \to C_0(\tilde{K}) \to 0 \to 0 \to .$$

Since $u: W \to \tilde{X}$ is a homotopy equivalence, we see that $u_*: C_*(W) \to C_*(\tilde{X})$ is a chain equivalence. Moreover it is a map of $Z\pi$-chain complexes.

It now follows that for any π-module $\underline{B}$,

$$H^{n+j}(X; \underline{B}) = H^{n+j}\left(\text{Hom}_\pi\left(X_*(\tilde{X}), \underline{B}\right)\right)$$

$$\simeq H^{n+j}\left(\text{Hom}_\pi\left(C_*(W), \underline{B}\right)\right)$$

$$= 0 \quad \text{if } j \ge 1.$$

Also $H_i(\tilde{X}) \simeq H_i(W) = 0$ for $i > n$. Thus X satisfies D_n.

Consider the inclusion $L = X^n \xrightarrow{\Psi} X$. It is an n-connected map. From Lemma 2.7 we see that X is dominated by L. Moreover the inclusion $K = K^{n-1} \xrightarrow{i} X$ is $(n-1)$-connected and $\pi_n(i) \simeq A$ [immediate from $\tilde{K} \vee (\vee_{\mu \in I} S^n_\mu) \to \tilde{X}$ being a homotopy equivalence]. By definition $\tilde{w}(X) = (-1)^n[A] = x$.

This proves Theorem 2.11. $\square$

Remark 2.12

We have seen already in Chapter 5 that there exist finitely presented groups π with $\tilde{K}_0(Z\pi) \neq 0$. We can pick a finite complex $K = K^2$ of dimension 2 with $\pi_1(K^2) \simeq \pi$. From Theorem 2.11, we have a finitely dominated CW-complex X with $X^2 = K$ and $\tilde{w}(X) = x \neq 0$ in $\tilde{K}_0(Z\pi)$. Thus a space dominated by a finite complex need not be of the homotopy type of a finite complex.

3   A PRODUCT FORMULA FOR WALL'S OBSTRUCTION

The material in this section is from Gersten's paper [7].

Throughout R will denote a ring. All the chain complexes we consider will be in R-mod and they will all be positive chain complexes.

Definition 3.1

A chain complex $\{C_n, d_n\}$ will be said to be of finite type if each C_n is f.g. over R.

   $C = \{C_n, d_n\}$ will be called a free chain complex if each C_n is free over R.
   $C = \{C_n, d_n\}$ will be called a finite projective complex (abbreviated as an FP complex) if each C_n is f.g. projective over R and $C_i = 0$ for $i > n_0$ for some n_0.
   $C = \{C_n, d_n\}$ will be called a finite free complex if each C_n is f.g.-free over R and $C_i = 0$ for $i > n_0$ for some n_0.

Proposition 3.2

Let $C = \{C_n, d_n\}$ and $C' = \{C'_n, d'_n\}$ be FP complexes and $f: C \to C'$ a chain map inducing isomorphisms in homology [i.e., $H_*(f): H_*(C) \simeq H_*(C')$]. Then

$$\sum_{i \geq 0} (-1)^i[C_i] = \sum_{i \geq 0} (-1)^i[C'_i] \quad \text{in } K_0(R). \tag{3.3}$$

[Observe that (3.3) is valid in $K_0(R)$ itself. We do not have to go to $\tilde{K}_0(R)$.]

Proof. Pick an n_0 with $C_i = 0 = C_i'$ for $i > n_0$. Let $\mathrm{CM}(f)$ denote the mapping cone of f. Observe that $\mathrm{CM}_n(f) = C_n' \oplus C_{n-1}$. Since $H_*(f)\colon H_*(C) \simeq H_*(C')$, we see that

$$0 \to 0 \to C_{n_0} \to C_{n_0}' \oplus C_{n_0-1} \to \cdots \to C_1' \oplus C_0 \to C_0' \to 0 \to 0 \to$$

is exact. This yields the equality

$$\sum_{i \geq 0} (-1)^i [C_i] = \sum_{i \geq 0} (-1)^i [C_i'] \quad \text{in } K_0(R). \qquad \square$$

Definition 3.4

Let $C = \{C_n, d_n\}$ be a free chain complex of finite type over R. We say that C *satisfies condition* G_N if

(i) $H_{N+i}(C) = 0$ for $i \geq 1$.

(ii) C_N/B_N is projective in R-mod where $B_N = \mathrm{Im}\, d_{N+1}\colon C_{N+1} \to C_N$.

Lemma 3.5

(i) *If C satisfies G_N, then C satisfies G_{N+i} for all $i \geq 0$.*

(ii) *Moreover $[C_N/B_N] = (-1)^i [C_{N+1}/B_{N+i}]$ in $\tilde{K}_0(R)$ for any $i \geq 0$.*

Proof.

(i) It suffices to show that $G_N \Rightarrow G_{N+1}$. If C satisfies G_N, then clearly $H_{N+1+i}(C) = 0$ for $i \geq 1$.

From $H_{N+1}(C) = 0$ we see that $Z_{N+1} = B_{N+1}$ where $Z_{N+1} = \ker d_{N+1}\colon C_{N+1} \to C_N$. Thus we get exact sequences

$$0 \to B_N \to C_N \to C_N/B_N \to 0, \tag{1}$$

$$0 \to B_{N+1} \to C_{N+1} \overset{\bar{d}}{\to} B_N \to 0, \tag{2}$$

where $\bar{d}\colon C_{N+1} \twoheadrightarrow B_N$ is induced by $d_{N+1}\colon C_{N+1} \to C_N$.

Since C_N/B_N is projective over R, (1) splits. Hence B_N is f.g. projective and (2) yields $C_N/B_{N+1} \simeq B_N$. Thus C_{N+1}/B_{N+1} is projective.

(ii) Since C_N and C_{N+1} are f.g.-free over R from (1) and (2) we get $[C_N|B_N] = -[B_N] = -[C_{N+1}/B_{N+1}]$ in $\tilde{K}_0(R)$. Now (ii) follows by induction on i. $\square$

Proposition 3.6

Let C satisfy G_N. Then C is chain equivalent to an FP complex over R.

Proof. Let $i_N: B_N \to C_N$ denote the inclusion. The map $d_{N+1}: C_{N+1} \to C_N$ factors as $C_{N+1} \xrightarrow{\bar{d}} B_N \xrightarrow{i_N} C_N$ with $\bar{d}: C_{N+1} \to B_N$ an epimorphism. Let P_* be the FP complex over R given by $P_i = C_i$ for $i \le N$, $P_{N+1} = B_N$ and $P_i = 0$ for $i > N + 1$. Let $f: C \to P_*$ be the chain map defined by $f_i = \text{Id}_{C_i}$ for $i \le N$ and $f_{N=1} = \bar{d}: C_{N+1} \to B_N = P_{N+1}$ $f_i = 0$ for $i > n + 1$. It is clear that f induces isomorphisms in homology. Since C and P_* are projective complexes over R, it follows that f is a chain equivalence. $\square$

Remark 3.7

Let C satisfy G_N. From Lemma 3.5 we see that for any $i \ge 0$ if we set $\tilde{w}(C) = (-1)^{N+i}[C_{N+i}/B_{N+i}] \in \tilde{K}_0(R)$, then $\tilde{w}(C)$ depends only on C and not on i.

Also from Proposition 3.6, we see that C is chain equivalent to the FP complex P_* given by $P_i = C_i$ for $i \le N$, $P_{N+1} = B_N$ with $P_i \xrightarrow{d_i} P_{i-1}$ the same as $d_i: C_i \to C_{i-1}$ for $i \le N$ and $P_{N+1} \to P_N$ the inclusion of B_N in P_N. Let $\eta: K_0(R) \to \tilde{K}_0(R)$ denote the quotient map. Then $\eta(\sum_{i \ge 0}(-1)^i[P_i]) = (-1)^{N+1}[B_N] = (-1)^N[C_N/B_N] = \tilde{w}(C) \in \tilde{K}_0(R)$. Also from Proposition 3.2 if P_*' is any FP-complex chain equivalent to C, then $\sum_{i \ge 0}(-1)[P_i] = \sum_{i \ge 0}(-1)^i[P_i']$ in $K_0(R)$. Hence

$$\eta\left(\sum_{i \ge 0}(-1)^i[P_i] \right) = \eta\left(\sum_{i \ge 0}(-1)^i[P_i'] \right) \quad \text{in } \tilde{K}_0(R).$$

Proposition 3.8

Let X be dominated by a finite complex. Let $\pi = \pi_1(X)$. Then the singular chain complex $CS(\tilde{X})$ of the universal covering of X is chain equivalent over $Z\pi$ to an FP complex P_*. Moreover the element $\sum(-1)^i[P_i]$ in $K_0(Z\pi)$ depends only on the homotopy type of X. Its image in $\tilde{K}_0(Z\pi)$ under the canonical quotient map $\eta: K_0(Z\pi) \to \tilde{K}_0(Z\pi)$ is the Wall obstruction $\tilde{w}(X)$ of X.

Proof. From Theorem 2.10 we see that $\exists$ an $n_0 \ge 3$ with X satisfying F_n and D_n for all $n \ge n_0$. In particular X satisfies F. From Theorem 1.12 we can assume that X itself is a CW-complex of finite type. Let $\mu: X^{n_0} \to X$ denote the inclusion. It is an n_0-connected map and $H_{n+1}(\tilde{\mu}) \simeq \pi_{n+1}(\tilde{\mu}) \simeq \pi_{n+1}(\mu)$ is a f.g. projective $Z\pi$-module. The singular chain complex $CS(\tilde{X})$ is $Z\pi$ chain equivalent to the cellular chain complex $C_*(\tilde{X})$. Since X is of finite type,

$C_*(\tilde{X})$ is a free $Z\pi$ chain complex of finite type. The cellular complex of the pair $(\tilde{X}, \tilde{X}^{n_0})$ is given by

$$\cdots \longrightarrow C_{n_0+3}(\tilde{X}) \xrightarrow{d_{n+3}} C_{n_0+2}(\tilde{X}) \xrightarrow{d_{n+2}} \left(C_{n_0+1}(\tilde{X}) \to 0 \cdots\right.$$

where $\{C_k(\tilde{X}), d_k\}_{k \geq 0} = C_*(\tilde{X})$ is the cellular chain complex of $\tilde{X}$. Since X satisfies D_{n_0}, $H_i(\tilde{X}) = 0$ for $i \geq n_0 + 1$. In particular we have

$$H_i(\tilde{X}, \tilde{X}^{n_0}) \simeq H_i(\tilde{X}) = 0 \quad \text{for } i \geq n_0 + 2.$$

Moreover

$$0 \to B_{n_0+1}(\tilde{X}) \to C_{n_0+1}(\tilde{X}) \to H_{n_0+1}(\tilde{X}, \tilde{X}^{n_0}) \to 0 \tag{$*$}$$

is exact in $Z\pi$-mod. Since $H_{n_0+1}(\tilde{X}, \tilde{X}^{n_0})$ is projective over $Z\pi$ the sequence $(*)$ splits.

$C_{n_0+1}(\tilde{X})|B_{n_0+1}(\tilde{X}) \simeq H_{n_0+1}(\tilde{X}, \tilde{X}^n)$ is $Z\pi$ projective. Thus the complex $C_*(\tilde{X})$ satisfies condition G_{n_0+1} of Gersten.

From Proposition 3.6 we see that $C_*(\tilde{X})$ is chain equivalent over $Z\pi$ to an FP complex P. From Remark 3.7 we see that the element $\Sigma(-1)^i[P_i] \in K_0(Z\pi)$ depends only on the homotopy type of X. Also $\eta(\Sigma(-1)^i[P_i] = (-1)^{n_0+1}[H_{n_0+1}(\tilde{X}, \tilde{X}^{n_0})] = (-1)^{n_0+1}[\pi_{n_0+1}(\mu)] = \tilde{w}(X)$. This proves Proposition 3.8. $\square$

Remark 3.9

When X is finitely dominated, with $\pi = \pi_1(X)$ we saw that $CS(\tilde{X})$ is chain equivalent over $Z\pi$ to an FP complex P_* and that $\Sigma(-1)^i[P_i] \in K_0(Z\pi)$ depends only on the homotopy type of X. Moreover if $\eta = K_0(Z\pi) \to \tilde{K}_0(Z\pi)$ denotes the quotient map, then $\eta(\Sigma(-1)^i[P_i]) = \tilde{w}(X)$, the obstruction to finiteness of X. Because of this reason, the element $\Sigma(-1)^i[P_i] \in K_0(Z\pi)$ is called the "unreduced Wall obstruction of X" and is denoted by $w(X)$. The element $\tilde{w}(X) \in \tilde{K}_0(Z\pi)$ will be called the reduced Wall obstruction of X.

Lemma 3.10

Let $0 \to C' \xrightarrow{j} C \xrightarrow{\varepsilon} C'' \to 0$ be an exact sequence of chain complexes over R and let $H_{N+1}(C'') = 0$. Let B'_N, B_N, and B''_N have the usual meanings ($B'_N = \text{Im } d'_{N+1}: C'_{N+1} \to C'_N$, etc.). Then

$$0 \longrightarrow B'_N \xrightarrow{j/B'_N} B_N \xrightarrow{\varepsilon/B_N} B''_N \longrightarrow 0 \quad \text{is exact.}$$

Proof. In fact $j: C_N' \to C_N$ is monic and hence its restriction $j|B_N'$ is monic. Let $u'' \in B_N''$. Then $u'' = d_{N+1}''c''$ for some $c'' \in C_{N+1}''$. Since $\varepsilon: C_{N+1} \to C_{N+1}''$ is onto, $\exists$ a $c \in C_{N+1}$ with $\varepsilon(c) = c''$. The commutativity of

$$
\begin{array}{ccc}
C_{N+1} & \xrightarrow{\ \varepsilon\ } & C_{N+1}'' \\
\downarrow{\scriptstyle d_{N+1}} & & \downarrow{\scriptstyle d_{N+1}''} \\
C_N & \xrightarrow{\ \varepsilon\ } & C_N''
\end{array}
$$

yields $\varepsilon d_{N+1}(c) = d_{N+1}''\varepsilon(c) = d_{N+1}''c'' = u''$. Now $u = d_{N+1}(c)$ is in B_N and $\varepsilon(u) = u''$. Thus $B_N \xrightarrow{\ \varepsilon|B_N\ } B_N'' \to 0$ is exact. From $e \circ j = 0$ we see that $\varepsilon|B_N \circ j|B_N' = \varepsilon \circ j|B_N' = 0$.

Let $x \in B_N$ satisfy $\varepsilon(x) = 0$. Since $x \in B_N$ we get an element $y \in C_{N+1}$ with $d_{N+1}y = x$. Then $\varepsilon y \in C_{N+1}''$ satisfies $d_{N+1}''(\varepsilon y) = \varepsilon d_{N+1}y = \varepsilon x = 0$. Since $H_{N+1}(C'') = 0$, $\exists$ an element $u'' \in C_{N+2}''$ with $\varepsilon y = d_{N+2}''u''$. Since $\varepsilon: C_{N+2} \to C_{N+2}''$ is onto, $\exists\ u \in C_{N+2}$ with $\varepsilon u = u''$. Now

$$
\begin{aligned}
\varepsilon(y - d_{N+2}u) &= \varepsilon y - \varepsilon d_{N+2}u \\
&= d_{N+2}''u'' - d_{N+2}''\varepsilon u = d_{N+2}''u'' - d_{N+2}''u'' = 0.
\end{aligned}
$$

Hence $\exists$ a $c' \in C_{N+1}'$ with $y - d_{N+2}u = j(c')$. Let $b' = d_{N+1}'c' \in B_N'$. Then

$$
\begin{aligned}
j(b') = jd_{N+1}'c' &= d_{N+1}j(c') = d_{N+1}(y - d_{N+2}u) \\
&= d_{N+1}y = x.
\end{aligned}
$$

This shows that $B_N' \xrightarrow{\ j|B_N'\ } B_N \xrightarrow{\ \varepsilon|B_N\ } B_N'$ is exact.

Thus combining everything so far proved we see that

$$
0 \longrightarrow B_N' \xrightarrow{\ j|B_N'\ } B_N \xrightarrow{\ \varepsilon|B_N\ } B_N'' \longrightarrow 0
$$

is exact. $\square$

Lemma 3.11

Let

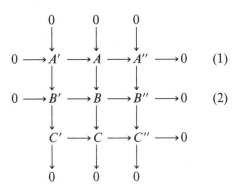

be a commutative diagram in R-mod with rows (1) *and* (2) *exact and all the columns exact. Then*

$$0 \to C' \to C \to C'' \to 0$$

is exact.

Proof. Standard. ☐

Theorem 3.12

Let $0 \to C' \overset{j}{\to} C \overset{\varepsilon}{\to} C'' \to 0$ *be an exact sequence of chain complexes in R-mod, each being a free chain complex of finite type. Suppose* C' *satisfies* $G_{N'}$ *and* C'' *satisfies* $G_{N''}$. *Then* C *satisfies* G_N *where* $N = \max(N', N'')$. *Moreover* $\tilde{w}(C) = \tilde{w}(C') + \tilde{w}(C'')$ *in* $\tilde{K}_0(R)$.

Proof. We have $H_{N=i}(C') = 0 = H_{N+i}(C'')$ for all $i \geq 1$ since $N = \max(N', N'')$. The exactness of $H_{N+1}(C') \overset{j_*}{\longrightarrow} H_{N+i}(C) \overset{\varepsilon_*}{\longrightarrow} H_{N+i}(C'')$ implies $H_{N+i}(C) = 0$ for $i \geq 1$. Lemma 3.10 implies that

$$0 \to B'_N \overset{j}{\to} B_N \overset{\varepsilon}{\to} B''_N \to 0$$

is exact. Lemma 3.11 now implies that

$$0 \to C'_N/B'_N \to C_N/B_N \to C''_N/B''_N \to 0 \tag{*}$$

is exact in R-mod. By definition $\tilde{w}(C') = (-1)^N[C'_N/B'_N] \in \tilde{K}_0(R)$, etc. The exactness of $(*)$ yields $[C_N/B_N] = [C'_N/B'_N] + [C''_N/B''_n]$ in $\tilde{K}_0(R)$ and hence $\tilde{w}(C) = \tilde{w}(C') + \tilde{w}(C'')$ in $\tilde{K}_0(R)$. ☐

Let π, π' be any two groups and $\Lambda = Z(\pi \times \pi')$ the group ring of $\pi \times \pi'$. Let C be a chain complex over $Z\pi$ and C' a chain complex over $Z\pi'$. Then the chain complex $C \otimes_Z C'$ naturally acquires the structure of a $\Lambda = Z(\pi \times \pi')$ chain complex. When A is a $Z\pi$-module and A' is a $Z\pi'$-module, then $\pi \times \pi'$ acts on $A \otimes_Z A'$ by

$$\sigma \times \sigma'(a \otimes a') = \sigma a \otimes \sigma' a' \qquad \forall \, \sigma \in \pi, \sigma' \in \pi'.$$

If A is f.g. over $Z\pi$ and A' is f.g.-free over $Z\pi'$, then $A \otimes_Z A'$ is f.g.-free over $Z(\pi \times \pi')$. In particular if C is a free chain complex of finite type over $Z\pi$ and C'' a free chain complex of finite type over $Z\pi'$, then $C \otimes_Z C'$ is a free chain complex of finite type over $Z(\pi \times \pi')$.

Lemma 3.13

Let C' be a chain complex over $Z\pi'$ satisfying $C_i' = 0$ for $i \neq r$ and C_r' f.g.-free over $Z\pi'$ of rank α_r. Let C be a free $Z\pi$-chain complex of finite type satisfying condition G_N. Then $C \otimes_Z C'$ satisfies condition G_{N+r} as a chain complex over $\Lambda = Z(\pi \times \pi')$. If $i: Z\pi \to \Lambda$ denotes the ring homomorphism induced by the inclusion $\pi \to \pi \times \pi'$, then

$$\tilde{w}(C \otimes_Z C') = (-1)^r \alpha_r \cdot i_*(\tilde{w}(C)) \quad in \ \tilde{K}_0(\Lambda).$$

Proof. Since C_r' is free over $Z\pi'$ and hence free over Z, we get

$$H_{j+r}(C \otimes_Z C') \simeq H_j(C) \otimes_Z C_r'.$$

Hence $H_{N+i+r}(C \otimes_Z C') \simeq H_{N+i}(C) \otimes_Z C_r' = 0$ for $i \geq 1$. Moreover $B_{N+r}(C \otimes_Z C') = B_N(C) \otimes_Z C_r'$ and

$$(C \otimes_Z C')_{N+r}/B_{N+r}(C \otimes_Z C') \simeq (C_N/B_N) \otimes_Z C_r'$$

$$\simeq \bigoplus_{\substack{\alpha_r \\ \text{copies}}} ((C_N/B_N) \otimes_Z Z\pi').$$

$(C_N/B_N) \otimes_Z Z\pi'$ is f.g. projective over Λ. Thus $C \otimes_Z C'$ satisfies G_{N+r} over Λ. Moreover $\tilde{w}(C \otimes_Z C') = (-1)^{N+r} \cdot \alpha_r[(C_N/B_N) \otimes_Z Z\pi'] \in \tilde{K}_0(\Lambda)$. Now

$$(C_N/B_N) \otimes_Z Z\pi' \simeq Z\pi' \otimes_Z (C_N/B_N) \quad in \ \Lambda\text{-mod}$$

$$\simeq Z\pi' \otimes_Z Z\pi \otimes_{Z\pi} (C_N/B_N)$$

$$\simeq \Lambda \otimes_{Z\pi} (C_N/B_N).$$

Hence $\tilde{w}(C \otimes_Z C') = (-1)^r \alpha_r (-1)^N [\Lambda \otimes_{Z\pi} C_N/B_N] = (-1)^r \alpha_r i_*(\tilde{w}(C))$. ☐

Theorem 3.14

Let C be a free $Z\pi$ complex of finite type satisfying condition G_N and let C' be a finite free complex over C'. Let r be an integer with $C_i' = 0$ for $i \geq r + 1$. Then $C \otimes_Z C'$ is a free $\Lambda = Z(\pi \times \pi')$ complex of finite type satisfying G_{N+r}. Moreover

$$\tilde{w}(C \otimes_Z C') = \chi(C') \cdot i_*(\tilde{w}(C))$$

where $i: Z\pi \to \Lambda = Z(\pi \times \pi')$ is induced by the inclusion $\pi \to \pi \times \pi'$.

Proof. The smallest integer r with $C_i' = 0$ for $i > r$ will be referred to as the dimension of C'. We prove the theorem by induction on r. When $r = 0$, this is immediate from Lemma 3.13. Assume the theorem there for complexes C' of dim $\leq r - 1$. Let now dim $C' = r$. Let $C'^{(r-1)}$ denote the $(r - 1)$ skeleton of C', namely $C_i'^{(r-1)} = C_i'$ for $i \leq r - 1$ and $C_i'^{(r-1)} = 0$ for $i \geq r$. The boundary maps $C_i'^{(r-1)} \to C_{i-1}'^{(r-1)}$ are the same as $d_i : C_i \to C_{i-1}$ for $i \leq r - 1$. Then $c'^{(r-1)}$ is a finite free $Z\pi'$ complex. The quotient complex $c'/C'^{(r-1)}$ is a free $Z\pi'$ complex with $(C'/C'^{(r-1)})_r = C_r'$ and $(C'/c'^{(r-1)})_j = 0$ for $j \neq r$. The exact sequence $0 \to C'^{(r-1)} \to C' \to C'/C'^{(r-1)} \to 0$ yields the exact sequence

$$0 \to C \otimes_Z C'^{(r-1)} \to C \otimes_Z C' \to C \otimes_Z \left(C'/C'^{(r-1)}\right) \to 0$$

of Λ complexes. By the inductive assumption $C \otimes_Z C'^{(r-1)}$ satisfies G_{N+r-1} and $\tilde{w}(C \otimes_Z C'^{(r-1)}) = \chi(C'^{(r-1)})i_*(\tilde{w}(C))$. Also by Lemma 3.13, $C \otimes_Z (C'/C'^{(r-1)})$ satisfies G_{N+r} and $\tilde{w}(C \otimes_Z (C'/C'^{(r-1)})) = (-1)^r \alpha_r i_*(\tilde{w}(C))$ where α_r equals the rank of C_r' over $Z\pi'$.

From Theorem 3.5 we now see that $C \otimes_Z C'$ satisfies G_{N+r} and

$$\tilde{w}(C \otimes_Z C') = \tilde{w}\left(C \otimes_Z C'^{(r-1)}\right) + \tilde{w}\left(C \otimes (C \otimes C'/C'^{(r-1)})\right)$$

$$= \chi(C'^{(r-1)})i_*(\tilde{w}(C)) + (-1)^r \alpha_r i_*(\tilde{w}(C))$$

$$= \chi(C')i_*(\tilde{w}(C)).$$

Theorem 3.15

Let X be dominated by a finite complex and Y be of the homotopy type of a finite complex. Let $i : \pi_1(X) \to \pi_1(X \times Y) = \pi_1(X) \times \pi_1(Y)$ denote the inclusion. Then $\tilde{w}(X \times Y) = \chi(Y)i_*(\tilde{w}(X)) \in \tilde{K}_0(Z\pi_1(X \times Y))$.

Proof. We can assume that X is a CW-complex of finite type and that Y is a finite CW-complex.

Then $C(\tilde{X})$ is a free $Z\pi_1(X)$ complex of finite type satisfying G_N for some N (≥ 3). $C(\tilde{Y})$ is a finite free $Z\pi_1(Y)$ complex. [Here $C(\tilde{X})$ and $C(\tilde{Y})$ are the cellular chain complexes of $\tilde{X}$ and $\tilde{Y}$, respectively.] Moreover $C(\tilde{X} \times \tilde{Y})$ can be identified with $C(\tilde{X}) \otimes_Z C(\tilde{Y})$. Theorem 3.15 is immediate from Theorem 3.14. □

Corollary 3.16

Let X be dominated by a finite complex. Then for any odd k, $X \times S^k$ is homotopy equivalent to a finite complex.

Chapter Seven

Finitely Dominated Nilpotent Spaces

In this chapter we present results about the Wall obstruction of finitely dominated nilpotent spaces. Most of these results are owing to Guido Mislin. All the spaces considered will be 0-connected and of the homotopy type of a CW-complex. As usual $\tilde{X}$ will denote the universal covering of X. The main references for the results in this chapter are [9, 13–15, and 23].

1 CONDITIONS FOR FINITE DOMINATION IN THE NILPOTENT CASE

In this section we will give a simple criterion for finite domination in the case of nilpotent spaces. But first we deal with a more general situation. Let $\pi = \pi_1(X)$. Let NF_2 and NF_n for $n \geq 3$ denote, respectively, the following conditions on the space X:

$(\mathrm{NF})_2$: π is finitely presented and $H_2(\tilde{X})$ is f.g. over $Z\pi$.

$(\mathrm{NF})_n$ $(n \geq 3)$: $(\mathrm{NF})_{n-1}$ holds and $H_n(\tilde{X})$ is f.g. over $Z\pi$.

Theorem 1.1

Let $Z\pi$ be noetherian. Then for $n \geq 2$ the conditions $(\mathrm{NF})_n$ and F_n (Chapter 6, Definition 1.7) are equivalent.

Proof. Let K be any finite complex with fundamental group π and $\varphi: K \to X$ be an $(n-1)$-connected map if $n \geq 3$ or a map inducing isomorphisms at the π_1 level if $n = 2$. Then $\pi_n(\varphi) \simeq H_n(\tilde{\varphi})$ and the sequence

$$H_n(\tilde{K}) \rightarrow H_n(\tilde{X}) \rightarrow H_n(\tilde{\varphi}) \rightarrow H_{n-1}(\tilde{K})$$

is exact. Since K is a finite complex, $H_{n-1}(\tilde{K})$ and $H_n(\tilde{K})$ are f.g. over $Z\pi$. Using the noetherian nature of $Z\pi$ we see that $H_n(\tilde{\varphi})$ is f.g. over $Z\pi$ if and only if $H_n(\tilde{X})$ is f.g. over $Z\pi$. □

We denote the class of nilpotent spaces by $\underline{\underline{NH}}$.

Theorem 1.2

Let $X \in \underline{\underline{NH}}$. Then the following are equivalent for X:

  (i) $\pi_n(X)$ is f.g. for all $n \geq 1$.
  (ii) $H_n(X)$ is f.g. for all $n \geq 1$.
  (iii) X is homotopically equivalent to a CW-complex of finite type.

Proof. We will accept the equivalence of (i) and (ii). Refer to ([8], Chapter 2, Theorem 2.16) for a proof of (i) ⟺ (ii). That (iii) ⟹ (i) is trivial.

We will prove (i) ⟹ (iii). Now $\tilde{X}$ is 1-connected with $\pi_n(\tilde{X}) \simeq \pi_n(X)$ all f.g. for $n \geq 2$. Hence it follows that $H_i(\tilde{X})$ are f.g. abelian groups for all i. In particular $H_i(\tilde{X})$ are all f.g. over $Z\pi$ for all i. Moreover π is a f.g. nilpotent group, hence π is finitely presented and $Z\pi$ is noetherian. From Theorem 1.1 we see that X satisfies all the F_n's. From Theorem 1.11 of Chapter 6 we see that X is homotopically equivalent to a CW-complex of finite type. □

Definition 1.3

A space X is said to be *quasi-finite* if $H_i(X)$ is f.g. for all i and there exists an integer ℓ with $H_i(X) = 0$ for $i > \ell$.

Lemma 1.4

Let $F \rightarrow X \rightarrow S^1$ be a fibration of nilpotent spaces with X quasi-finite. Then F is quasi-finite.

Proof. In the Serre spectral sequence of this fibration, we have $E^2_{p,q} = H_p(S^1, H_q(F))$, pth homology of S^1 with coefficients in the local system $H_q(F)$. Thus $E^2_{p,q} = 0$ for $p \geq 2$ and also for $p < 0$ or $q < 0$. It follows that $E^2_{0,q} = E^3_{0,q} = \cdots = E^\infty_{0,q}$. But $E^\infty_{0,q} = F_0 H_q(X)$. Also $E^2_{1,q-1} = \cdots =$

$E^\infty_{1,1-1}$. Hence

$$\frac{F_1 H_q(X)}{F_0 H_q(X)} = E^\infty_{1,q-1} = \cdots = E^2_{1,1-1} = H_1\big(S^1, H_{q-1}(F)\big).$$

Also $E^2_{p,q} = 0$ for $p \geq 2 \Rightarrow E^\infty_{p,q} = 0$ for $p \geq 2$. In particular

$$\frac{F_2 H_q(X)}{F_1 H_q(X)} = E^\infty_{2,q-2} = 0, \qquad \frac{F_3 H_q(X)}{F_2 H_q(X)} = 0, \quad \text{etc.}$$

$$\frac{F_q H_q(X)}{F_{q-1} H_q(X)} = 0, \qquad F_q H_q(X) = H_q(X).$$

Thus $F_1 H_q(X) = H_q(X)$. From the exactness of

$$0 \to F_0 H_q(X) \to F_1 H_q(X) \to F_1 H_q(X)/F_0 H_q(X) \to 0$$

we see that

$$0 \to H_0\big(S^1; H_q(F)\big) \to H_q(X) \to H_1\big(S^1; H_{q-1}(F)\big) \to 0$$

is exact. But $H_0(S^1; H_q(F)) = H_q(F)/I\pi H_q(F)$ where $\pi = \pi_1(S^1)$ is the infinite cyclic group. It is known that the action of π on $H_q(F)$ is nilpotent ([9], Chapter 2, Proposition 6.1). Hence $(I\pi)^k H_q(F) = 0$ for some k (depending on q). Since X is quasi-finite, $H_q(X) = 0$ for $q \geq \ell$. Hence $H_q(F)/I\pi H_q(F) \simeq H_0(S^1; H_q(F)) = 0$ for $q \geq \ell$. This yields $H_q(F) = I\pi H_q(F)$. Iteration yields $H_q(F) = I\pi H_q(F) = (I\pi)^2 H_q(F) = \cdots = (I\pi)^k H_q(F) = 0$. Hence $H_q(F) = 0$ for $q \geq \ell$. From the homotopy exact sequence of the fibration $F \to X \to S^1$ we get $\pi_j(F) \simeq \pi_j(X)$ for $j \geq 2$ and $\{1\} \to \pi_1(F) \to \pi_1(X) \to Z \to \{1\}$ exact. Thus $\pi_1(F)$ is a subgroup of the finitely generated nilpotent group $\pi_1(X)$ and $\pi_j(F) \simeq \pi_j(X)$ for $j \geq 2$. Hence all the homotopy groups of F are f.g. Since F is nilpotent by [9], Chapter 2, Theorem 2.16, we see that $H_j(F)$ are all f.g. $\square$

Lemma 1.5

Let p be a prime and $F \to X \to K(Z/p, 1)$ a fibration of nilpotent spaces with X quasi-finite. Then F is quasi-finite.

Proof. From the homotopy exact sequence of the fibration $F \to X \to K(Z/p, 1)$ we see that the homotopy groups of F are all f.g. Using (i) $\Leftrightarrow$ (ii) of Theorem 1.2, we have only to prove that $H_i(F; Z) = 0$ for $i \geq \ell$, for a suitable integer ℓ. In fact it suffices to show that $H_i(F; Z/q) = 0$ for $i \geq \ell$ for all primes q. Let n be such that $H_i(X; Z/q) = 0$ for $i \geq n$. If q is a prime not equal to p, since $F \to X$ is a q equivalence, it follows that $H_i(F; Z/q) = 0$ for $i \geq n$. Let $q = p$. $F \to X$ can be regarded as the pullback of the universal covering $T \to K(Z/p, 1)$ of $K(Z/p, 1)$. Thus $H_i(X; Z/p) \simeq H_i(CS(F) \otimes_\pi Z/p)$. Observe that $\pi = Z/p$ acts on F. Also

$$H_i(F; Z/p) \simeq H_i(CS(F) \otimes_\pi Z/p[\pi]).$$

If ω generates the cyclic group, then $I\pi$ is generated by $\omega - 1$ over $Z/p[\pi]$. Moreover $(\omega - 1)^p = \omega^p - 1 = 0$ in $Z/p[\pi]$. Thus $(I\pi)^p = 0$. Thus $Z/p[\pi] \supset I\pi \supset (I\pi)^2 \supset \cdots (I\pi)^{p-1} \supset (I\pi)^p = 0$ is a filtration of $Z/p[\pi]$ with the graded associated object a Z/p vector space with trivial π action. At each stage the consecutive quotient is of dimension 1 over Z/p. Hence

$$\dim H_i(F; Z/p) \leq p \dim H_i(X; Z/p).$$

In particular $H_i(F; Z/p) = 0$ for $i \geq n$. This proves that F is quasi-finite. $\square$

If H is a subgroup of a finitely generated nilpotent group G, then there exists a normal chain

$$H = G_0 \subset G_1 \subset \cdots \subset G_n = G$$

with G_i/G_{i-1} either infinite cyclic or cyclic of prime order. The fundamental group of a quasi-finite nilpotent space is f.g. [by (i) $\Leftrightarrow$ (ii) of Theorem 1.2]. Now Lemmas 1.4 and 1.5 together yield the following theorem.

Theorem 1.6

Let $F \to X$ be a covering space of the quasi-finite nilpotent space X. Then F is quasi-finite nilpotent.

Lemma 1.7

Let π be a finite p group and M a f.g. artinian π-module. Then there is an exact sequence

$$0 \to P_2 \to P_1 \to M \to N \to 0$$

with N a nilpotent π-module and P_1, P_2 finitely generated projective π-modules.

Proof. Since M is artinian, one has $I^k M = I^{k+1} M$ for some integer $k \geq 1$, I being the augmentation ideal of $Z\pi$. By a result of Brown and Dror ([3], Theorem 3) the natural quotient map $M \to M/I^k M$ induces an isomorphism

$H_*(\pi; M) \simeq H_*(\pi; M/I^k M)$. It follows that $H_i(\pi; I^k M) = 0$ for all i. Hence there are integers (possibly negative) with $\hat{H}^j(\pi; I^k M) = \hat{H}^{j+1}(\pi; I^k M) = 0$. π is its own Sylow p group. From Theorem 4.9 of Chapter 5 we see that proj $\dim_{Z\pi} I^k M \leq 1$. Also $Z\pi$ is noetherian. Hence $I^k M$ is f.g. over $Z\pi$. We can get an exact sequence $0 \to P_2 \to P_1 \to I^k M \to 0$ with P_1, P_2 f.g. projective. Clearly $I^k(M/I^k M) = 0$. Hence $N = M/I^k M$ is nilpotent and $0 \to P_2 \to P_1 \to M \to M/I^k M \to 0$ is exact. $\square$

Lemma 1.8

Let X be a nilpotent space homotopically equivalent to a CW-complex of finite type with $\pi_1(X)$ a finite p group. Suppose there exists an integer $n \geq 2$ such that $H^{n+i}(X; Z) = 0 = H^{n+i}(\tilde{X}, Z)$ for all $i \geq 1$, where $\tilde{X}$ is the universal covering of X. Then $H^{n+i}(X; \mathscr{B}) = 0$ for all $i \geq 1$ and all local coefficient systems $\mathscr{B}$ on X.

Proof. Let M be any local coefficient system for cohomology over X [i.e., M is a left $Z\pi$-module where $\pi = \pi_1(X)$]. Then $H^i(\tilde{X}, M) = H^i(\text{Hom}_\pi(\text{CS}(\tilde{X}), M))$ where $\text{CS}(\tilde{X})$ is the singular chain complex of $\tilde{X}$. Since X is homotopically equivalent to a CW-complex of finite type, $\text{CS}(\tilde{X})$ is chain equivalent to a chain complex C_* over $Z\pi$ which is f.g.-free in each dimension. Hence $\varinjlim H^k(X, M_\alpha) \simeq H^k(X, \varinjlim M_\alpha)$ for any direct system $\{M_\alpha\}_{\alpha \in \Omega}$ of π-modules. Hence to prove that $H^{n+i}(X, M) = 0$ for all $i \geq 1$ and all π-modules M, we have only to show that $H^{n+i}(X, M) = 0$ for all $i \geq 1$ and all f.g. π-modules M. Since C_i is free f.g. over $Z\pi$ and π is finite, it follows that $\text{Hom}_\pi(C_i, M)$ is a f.g. abelian group whenever M is a f.g. $Z\pi$-module. Thus $\text{Hom}_\pi(C_*, M)$ is a cochain complex over Z with each cochain module f.g. over Z. Now, $M \otimes_Z (Z/q)$ can be regraded as a π-module via $\sigma(x \otimes_Z \lambda) = \sigma x \otimes_Z \lambda$ with $\sigma \in \pi$, $\lambda \in Z/q$, and $x \in M$. From $\text{Hom}_\pi(Z\pi, M \otimes Z/q) \simeq M \otimes Z/q$ and $\text{Hom}_\pi(Z\pi, M) \simeq M$ and the fact that C_i is f.g.-free over $Z\pi$ we get $\text{Hom}_\pi(C_*, M) \otimes Z/q \simeq \text{Hom}_\pi(C_*, M \otimes Z/q)$. Since $\text{Hom}_\pi(C_*, M)$ is a cochain complex of f.g. abelian groups, to prove that $H^{n+i}(C_*, M) = 0$ for $i \geq 1$, it suffices to show that $H^{n+i}(\text{Hom}_\pi(C_*, M) \otimes Z/q) = 0$ for all primes q. Thus we have only to show that $H^{n+i}(\text{Hom}_\pi(C_*, M \otimes Z/q)) = 0$ for every f.g. $M \in Z\pi$-mod. When M is f.g., $M \otimes_Z Z/q$ is artinian over $Z\pi$. Hence by Lemma 1.7, there exists an exact sequence $0 \to P_2 \to P_1 \to M \to N \to 0$ with N a f.g. nilpotent π-module and P_1, P_2 f.g. projective $Z\pi$-modules. From the exact sequences

$$0 \to P_2 \to P_1 \to P_1/P_2 \to 0,$$

$$0 \to P_1/P_2 \to M \to N \to 0,$$

we see that we need only prove that $H^{n+i}(X; P) = 0$ for $i \geq 1$ for any f.g.

projective $Z\pi$-module and $H^{n+i}(X; N) = 0$ for any f.g. nilpotent π-module N. Since P is a direct summand of a f.g.-free $Z\pi$-module, the first is proved if we show that $H^{n+i}(X; (Z\pi)^k) = 0$ for all $i \geq 1$. But $\mathrm{Hom}_\pi(\mathrm{CS}(\tilde{X}), (Z\pi^k)) \simeq \mathrm{Hom}_Z(\mathrm{CS}(\tilde{X}); Z)$. Hence $H^{n+i}(X; (Z\pi)^k) \simeq H^{n+i}(\tilde{X}; Z) = 0$ for $i \geq 1$. Suppose N is f.g. nilpotent over $Z\pi$. Using a filtration such that successive quotients are trivial π-modules (i.e., abelian groups with trivial π action) we have only to show that $H^{n+i}(X; A) = 0$ where A is a f.g. abelian group with trivial π action. But this follows from $H^{n+i}(X; Z) = 0$. $\square$

Proposition 1.9

Let X be a nilpotent space with $\pi_1(X)$ a finite p group. Then X is finitely dominated if and only if X is quasi-finite.

Proof. The universal covering $\tilde{X}$ is quasi-finite by Theorem 1.6. Thus there exists an integer N with $H^{N+i}(X; Z) = 0 = H^{N+i}(\tilde{X}; Z)$ for all $i \geq 1$. By Lemma 1.8, $H^{N+i}(X, \mathscr{B}) = 0$ for $i \geq 1$ and any local system $\mathscr{B}$ over X. From Theorem 1.2 we see that X satisfies F_n for all $n \geq 1$. Hence from Theorem 2.10 of Chapter 6 we see that X is dominated by a finite complex. $\square$

Let $\pi = \pi_1(X)$ and $\bar{\pi}$ be a subgroup. Let B be a local coefficient system over X, i.e., a π-module. Let $\tilde{X}$ denote the universal covering of X and $\hat{X}$ be the covering of X corresponding to the subgroup $\bar{\pi}$ of π. Then $\tilde{X}$ is also the universal covering of $\hat{X}$. Using the inclusion $\bar{\pi} \subset \pi$ we regard B as a $\bar{\pi}$-module (i.e., a local coefficient system over $\hat{X}$). The cohomology of $\mathrm{Hom}_\pi(\mathrm{CS}(\tilde{X}), B)$ is the cohomology of X with coefficients in the local system B, written $H^*(X; B)$. The cohomology of $\hat{X}$ with coefficients in the local system B is given by $H^*(\hat{X}; B) = H^*(\mathrm{Hom}_{\bar{\pi}}(\mathrm{CS}(\tilde{X}); B))$. The inclusion $\mathrm{Hom}_\pi(\mathrm{CS}(\tilde{X}); B) \to \mathrm{Hom}_{\bar{\pi}}(\mathrm{CS}(\tilde{X}); B)$ induces a map

$$i(\bar{\pi}, \pi): H^*(X; B) \to H^*(\hat{X}; B).$$

In case $\bar{\pi}$ is of finite index in π one defines the so-called transfer map

$$t(\pi, \bar{\pi}): H^*(\hat{X}; B) \to H^*(X; B)$$

as follows. Let $x_1, \ldots, x_n$ be a complete set of representatives for the left cosets of $\bar{\pi}$ in π. Let $\lambda: \mathrm{Hom}_{\bar{\pi}}(\mathrm{CS}(\tilde{X}); B) \to \mathrm{Hom}_\pi(\mathrm{CS}(\tilde{X}), B)$ be defined by

$$(\lambda f)(c) = \sum_{i=1}^{n} x_i f(x_i^{-1} c)$$

for any $f \in \operatorname{Hom}_{\bar{\pi}}(\operatorname{CS}_i(\tilde{X}); B)$, $c \in \operatorname{CS}_i(\tilde{X})$. λf is a π homomorphism. In fact if $\sigma \in \pi$ we have

$$(\lambda f)(\sigma \cdot c) = \sum_{i=1}^{n} x_i f(x_i^{-1}\sigma c)$$

$$= \sum_{i=1}^{n} \sigma \cdot \sigma^{-1} x_i f(x_i^{-1}\sigma c)$$

$$= \sigma \sum_{i=1}^{n} y_i f(y_i^{-1}c)$$

where $y_i = \sigma^{-1}x_i$.

Clearly $y_1, \ldots, y_n$ are also representatives for the distinct left cosets of $\bar{\pi}$ in π.

We now show that $\sum_{i=1}^{n} y_i f(y_i^{-1}c) = \sum_{i=1}^{n} x_i f(x_i^{-1}c)$. There exists a permutation θ of $\{1, \ldots, n\}$ and elements $\tau_i \in \bar{\pi}$ with $y_i = x_{\theta(i)}\tau_i$. Hence $f(y_i^{-1}x_{\theta(i)}c) = \tau_i^{-1}f(x_{\theta(i)}c)$. Hence

$$\sum_{i=1}^{n} y_i f(y_i^{-1}c) = \sum_{i=1}^{n} x_{\theta(i)}\tau_i\tau_i^{-1}f(x_{\theta(i)}c)$$

$$= \sum_{i=1}^{n} x_{\theta(i)} f(x_{\theta(i)}c)$$

$$= \sum_{i=1}^{n} x_i f(x_i^{-1}c).$$

Thus $(\lambda f)(\sigma c) = \sigma(\lambda f)(c) \ \forall \ \sigma \in \pi$. It is easy to check that λ is a map of cochain complexes. The induced map

$$t(\pi, \bar{\pi}): H^*(\hat{X}; B) \to H^*(X; B)$$

is called the transfer map. It is easily checked that $t(\pi, \bar{\pi})i(\bar{\pi}, \pi)$ equals multiplication by the index of $\bar{\pi}$ in π. $\square$

Lemma 1.10

Let X be a connected complex with finite fundamental group $\pi_1 X = \pi$. Let $X(p)$ denote a covering space of X associated with a p-Sylow subgroup $\pi(p)$ of π.

Then

$$i(\pi(p), \pi): H^*(X; B/pB) \to H^*(X(p); B/pB)$$

is monic for every π-module B.

Proof. The groups $H^*(X; B/pB)$ are vector spaces over Z/pZ and $t(\pi, \pi(p))i(\pi(p), \pi)$ is multiplication by $[\pi : \pi(p)]$ which is prime to p. Hence the result. $\square$

Proposition 1.11

Suppose X is a quasi-finite nilpotent space with finite fundamental group. Then X is finitely dominated.

Proof. Without loss of generality, we can assume X to be a CW-complex of finite type. Let $X(p)$ be the covering space of X corresponding to the p-Sylow subgroup $\pi(p)$ of $\pi_1(X) = \pi$. Then $X(p)$ is quasi-finite by Theorem 1.6 and $X(p) \sim \tilde{X}$ for almost all primes p. From Proposition 1.9 we see that $X(p)$ is finitely dominated for all primes p. Since $X(p) \sim \tilde{X}$ for almost all primes p, we see that there exists an integer N such that $H^i(X(p); C) = 0$ for $i > N$ and all $\pi(p)$-modules C. Let B be any f.g. π-module. Since $i(\pi(p): \pi)H^*(X; B/pB) \to H^*(X(p); B/pB)$ is monic by Lemma 1.10, we get $H^i(X; B/pB) = 0$ for $i > N$. Since X is of finite type, $H^i(X; B)$ is a f.g. abelian group. From the arguments used in the proof of Lemma 1.8 we see that $H^i(X; B/pB) = 0$ for $i > N$ and every prime $p \Rightarrow H^i(X; B) = 0$ for $i > N$. If D is an arbitrary π-module, then $D = \lim B_\alpha$ with B_α f.g. in $Z\pi$-mod. Since X is of finite type $H^i(X; D) = \lim H^i(X; B_\alpha) = 0$ for $i > N$. From Theorem 2.10 of Chapter 6 we see that $\vec{X}$ is dominated by a finite complex. $\square$

The following result is due to Lal [10]. Although some of the other results in that paper are not correct, the result quoted next is correct.

Theorem 1.12

Let $F \xrightarrow{i} E \xrightarrow{p} B$ be a fibration with F a finitely dominated complex and B a finite complex. Then E is finitely dominated and $w(E) = j_*(w(F))\chi(B)$ where w denotes the unreduced Wall obstruction and $j_*: \pi_1(F) \to \pi_1(E)$ the map induced by j.

Proposition 1.13

Let G be a torsion-free f.g. nilpotent group. Then $K(G,1)$ is of the homotopy type of a finite complex.

Proof (By Induction on the Nilpotency Index of G). If G is abelian, then $G \simeq Z^{\ell}$ for some ℓ and $K(Z,1) \sim S^1$. Let nil $G = k > 1$. If $Z(G)$ is the center of G, then $Z(G)$ is f.g. torsion-free abelian and $G/Z(G)$ is torsion-free of nilpotency index less than nilpotency index of G. By the inductive assumption $K(G/Z(G),1)$ is of the homotopy type of a finite complex. From the fibration $K(Z(G),1) \to K(G,1) \to K(G/Z(G),1)$ and Theorem 1.12 we get $K(G,1)$ to be finitely dominated. Thus for any torsion-free f.g. nilpotent group G, $K(G,1)$ is finitely dominated.

Next we show that $w(K(G,1)) = 0$. In fact there exists an epimorphism $G \xrightarrow{\eta} Z$ and ker $\eta = H$ is f.g. torsion-free nilpotent. We have a fibration $K(H,1) \to K(G,1) \to S^1$ and $K(H,1)$ is finitely dominated. Hence by Theorem 1.12,

$$w(K(G,1)) = j_*(w(K(H,1)))\chi(S^1) = 0. \qquad \square$$

Theorem 1.14

Let X be a nilpotent space. Then X is finitely dominated if and only if X is quasi-finite.

Proof. If X is quasi-finite nilpotent, then $\pi_1(X)$ is f.g. nilpotent. If $\pi_1(X)$ is a torsion group, then $\pi_1(X)$ is finite. Then the theorem is an immediate consequence of Proposition 1.11.

Suppose $\pi_1(X)$ is not a torsion group. Let T be the torsion subgroup. There T is a finite normal subgroup of $\pi_1(X)$ and $G = \pi_1(X)/T$ is a f.g. torsion-free nilpotent group. By Proposition 1.13, $K(G,1)$ is homotopically equivalent to a finite complex. If $\tilde{X}$ is the universal covering of X and $\tilde{X}/T$ is the quotient of $\tilde{X}$ under the action of T, then there exists a fibration $\tilde{X}/T \to X \to K(G,1)$. From Theorem 1.6 we see that $\tilde{X}/T$ is quasi-finite nilpotent. Also $\pi_1(\tilde{X}/T) = T$ is a finite nilpotent group. From Proposition 1.11 we see that $\tilde{X}/T$ is finitely dominated. From Theorem 1.12 we see that X is finitely dominated. $\square$

Theorem 1.15

Let X be a finitely dominated nilpotent space with infinite fundamental group. Then $w(X) = 0$. In particular X is homotopically equivalent to a finite complex.

Proof. $\pi_1(X)$ is a f.g. infinite nilpotent group. Hence there exists a surjection $\pi_1(X) \overset{\varphi}{\to} Z$. Let F be the covering of X associated to $\ker \varphi$. Then there exists a fibration $F \to X \to S^1$. F is quasi-finite nilpotent by Theorem 1.6. Hence F is finitely dominated by Theorem 1.14. From Theorem 1.12 we now see that X is finitely dominated and that $w(X) = j_*(\omega(F))\chi(S^1) = 0$. $\quad\square$

2 FINITELY DOMINATED NILPOTENT SPACES WITH A FINITE FUNDAMENTAL GROUP

In Section 1 we saw that if X is a finitely dominated nilpotent space with $\pi_1(X)$ *infinite*, then $w(X) = 0$ and hence X is homotopically equivalent to a finite complex. In the present section we will present some results on finitely dominated nilpotent spaces with a finite fundamental group.

For this purpose we introduce the concept of a special finite projective (SFP) complex over $Z\pi$ where π is any group. Let $Z\pi$-mod (resp. mod-$Z\pi$) denote the category of left (resp. right) $Z\pi$-modules. Take $Z \in \text{mod-}Z\pi$ with trivial π operators. For any $M \in Z\pi$-mod and any two subgroups $G \subset G'$ of π the canonical quotient map $Z \otimes_G M \to Z \otimes_{G'} M$ will be denoted by $\eta^M_{G,G'}$. If C is a chain complex in $Z\pi$-mod, there is an epimorphism $\eta^C_{G,G'}: Z \otimes_G C \to Z \otimes_{G'} C$ of chain complexes. When there is no possibility of confusion we merely write $\eta_{G,G'}$ for $\eta^M_{G,G'}$ or $\eta^C_{G,G'}$.

Definition 2.1

Let Γ denote a set of primes. A homomorphism $f: A \to B$ of abelian groups is called a Γ isomorphism if $\ker f$ and $\operatorname{coker} f$ are torsion groups with no p torsion for any $p \in \Gamma$.

Given any integer $k \geq 1$, let $\Gamma(k)$ denote the set of primes not dividing k.

Definition 2.2

An FP-complex C over $Z\pi$ is said to be *special* if it satisfies the following condition (S).

(S): For any two subgroups $G \subset G'$ of π with index $[G':G]$ of G in G' finite, the map

$$(\eta_{G,G'})_*: H_*(Z \otimes_G C) \to H_*(Z \otimes_{G'} C)$$

in homology induced by $\eta_{G,G'}$ is a $\Gamma([G':G])$ isomorphism.

In other words if $[G' : G] = k < \infty$, then the kernel and cokernel of $(\eta_{G,G'})_*$ are torsion groups with no p torsion for any p not dividing k.

Any FP-complex over $Z\pi$ satisfying condition (S) will be referred to as an *SFP complex*.

We need the following result from the theory of nilpotent spaces [9].

Lemma 2.3

Let $\varphi: Y \to X$ be a covering projection of nilpotent spaces with $\varphi_*(\pi_1(Y))$ of finite index k in $\pi_1(X)$. Then $\varphi_*: H_*(Y; Z) \to H_*(X; Z)$ is a $\Gamma(k)$ isomorphism.

Definition 2.4

A space X is said to be of type FP if the singular chain complex $\mathrm{CS}(\tilde{X})$ of the universal covering of X is chain equivalent to an FP-complex over $Z\pi$ where $\pi = \pi_1(X)$.

Proposition 2.5

Suppose X is a nilpotent space of type FP. Then any FP chain complex C chain equivalent to $\mathrm{CS}(\tilde{X})$ over $Z\pi$ is an SFP chain complex over $Z\pi$ where $\pi = \pi_1(X)$.

Proof. Let $G \subset G'$ be any two subgroups of π with $[G' : G] = k < \infty$. The covering projection $p: \tilde{X} \to X$ can be factored as $\tilde{X} \xrightarrow{g} Y \xrightarrow{\varphi} Y' \xrightarrow{h} X$ with g, φ, h all covering projections and satisfying $h_*(\pi_1(Y')) = G'$ and $(h \circ \varphi)_*(\pi_1(Y)) = G$. Choose a chain equivalence $\alpha: C \to \mathrm{CS}(\tilde{X})$ over $Z\pi$. Then α is automatically a chain equivalence over ZG and ZG'. Moreover $Z \otimes_G \mathrm{CS}(\tilde{X})$ and $Z \otimes_{G'} \mathrm{CS}(\tilde{X})$ can be identified, respectively, with $\mathrm{CS}(Y)$ and $\mathrm{CS}(Y')$ in such a way that the following diagram commutes.

$$
\begin{array}{ccc}
Z \otimes_G C & \xrightarrow{1 \otimes_G \alpha} & Z \otimes_G \mathrm{CS}(X) == \mathrm{CS}(Y) \\
\downarrow{\scriptstyle \eta_{G,G'}} & \downarrow{\scriptstyle \eta_{G,G'}} & \downarrow{\scriptstyle \mathrm{CS}(\varphi)} \\
Z \otimes_{G'} C & \xrightarrow{1 \otimes_{G'} \alpha} & Z \otimes_{G'} \mathrm{CS}(\tilde{X}) == \mathrm{CS}(Y')
\end{array}
$$

From Lemma 2.3 we see immediately that $(\eta_{G,G'})_*: H_*(Z \otimes_G C) \to H_*(Z \otimes_{G'} C)$ is a $\Gamma(k)$ isomorphism. $\square$

Let $\varphi: \pi \to \pi'$ be an epimorphism of groups, H a subgroup of π', and $G = \varphi^{-1}(H)$. Using the ring homomorphism $\varphi: Z\pi \to Z\pi'$ induced by φ, we regard $Z\pi'$ as an object of mod-$Z\pi$. Let $M \in Z\pi$-mod. Regard Z as an object of mod-$Z\pi'$ as well as of mod-$Z\pi$ with trivial action. Under these conditions the following lemma is easy to prove.

Lemma 2.6

There exist well-defined maps (as abelian groups)

$$\Psi_M^H: Z \otimes_G M \to Z \otimes_H (Z\pi' \otimes_\pi M),$$

$$\Phi_M^H: Z \otimes_H (Z\pi' \otimes_\pi M) \to Z \otimes_G M,$$

satisfying

$$\Psi_M^H (n \otimes_G u) = n \otimes_H (e' \otimes_\pi u),$$

$$\Phi_M^H (n \otimes_H (x' \otimes_\pi u)) = n \otimes_G xu,$$

for any $n \in Z$, $u \in M$ and $x' \in \pi'$ where e' is the identity element of π' and $x \in \pi$ any element satisfying $\varphi(x) = x'$. Moreover Φ_M^H, Ψ_M^H are natural in M and are isomorphisms, each being the inverse of the other.

Proposition 2.7

Let $\varphi: \pi \to \pi'$ be an epimorphism of groups and C an SFP complex over $Z\pi$. Then $Z\pi' \otimes_\pi C$ is an SFP complex over $Z\pi'$.

Proof. Write E for the complex $Z\pi' \otimes_\pi C$. It is clearly an FP-complex over $Z\pi'$. Let $H \subset H'$ be subgroups of π' with $[H' : H] = k < \infty$. Then $G = \varphi^{-1}(H)$ and $G' = \varphi^{-1}(H')$ are subgroups of π with $G \subset G'$ and $[G' : G] = k$. Also

$$
\begin{array}{ccc}
Z \otimes_G C & \xrightarrow{\Psi^H} & Z \otimes_H E \\
{\scriptstyle \eta_{G,G'}^C} \downarrow & & \downarrow {\scriptstyle \eta_{H,H'}^E} \\
Z \otimes_{G'} C & \xrightarrow{\Psi^{H'}} & Z \otimes_{H'} E
\end{array}
$$

is a commutative diagram. The horizontal maps are isomorphisms of complexes over Z. Hence $(\eta_{H,H'}^E)_*: H_*(Z \otimes_H E) \to H_*(Z \otimes_{H'} E)$ can be identified with $(\eta_{G,G'}^C)_*: H_*(Z \otimes_G C) \to H_*(Z \otimes_{G'} C)$. Since C is an SFP complex over $Z\pi$, it follows that $(\eta_{G,G'}^C)_*$ and hence $(\eta_{H,H'}^E)_*$ is a $\Gamma(k)$ isomorphism. $\square$

Proposition 2.8

Let π be any finite p group where p is a prime. Let C be an SFP *complex over $Z\pi$. Then $H_*(C)$ is a nilpotent π-module.*

Proof. Since $Z\pi$ is noetherian and C is a complex of finite length with each C_i finitely generated, it follows that $H_*(C)$ is finitely generated over $Z\pi$. Since π is finite, we see that $H_*(C)$ is finitely generated over Z. The complex $Z \otimes_\pi C$ is a complex of finite length with each $Z \otimes_\pi C_i$ finitely generated over Z. Hence $H_*(Z \otimes_\pi C)$ is f.g. over Z. Let e denote the unit element of π. It is easy to see that the map $\eta_{\{e\},\pi}: Z \otimes_{\{e\}} C = C \to Z \otimes_\pi C$ is the same as $\varepsilon \otimes_\pi 1:$ $Z\pi \otimes_\pi C = C \to Z \otimes_\pi C$ where $\varepsilon: Z\pi \to Z$ is the augmentation map. Write ε' for $\varepsilon \otimes_\pi 1$. Since $\varepsilon: Z\pi \to Z$ is a left π homomorphism with trivial left π action on Z, it follows that $\varepsilon'_*: H_*(C) \to H_*(Z \otimes_\pi C)$ is a left π homomorphism with trivial left π action on $H_*(Z \otimes_\pi C)$. Let $L = \text{Im } \varepsilon'_*$. Then $I\pi L = 0$. Since C is SFP and $[\pi:\{e\}] = p^k$, a power of p, it follows that $\ker \varepsilon'_*$ and coker ε'_* are p-primary torsion abelian groups. Since $H_*(C)$ is f.g. over Z we see that $K = \ker \varepsilon'_*$ is a finite abelian p group. It is well known that a finite p group can act only in a nilpotent way on another finite p group. Hence $(I\pi)^\ell K = 0$ for some ℓ. The exactness of $0 \to K \to H_*(C) \to L \to 0$ now yields $(I\pi)^{\ell+1}H_*(C) = 0$. $\square$

Lemma 2.9

Let π be any group and π' a subgroup of π of finite index and C an SFP *complex over $Z\pi$. Then C is an* SFP *complex over $Z\pi'$.*

Theorem 2.10

Let X be a space with $\pi_1(X)$ finite. Suppose $\text{CS}(\tilde{X})$ is chain equivalent over $Z\pi_1(X)$ to an SFP *complex. Then for any prime p, any p subgroup π' of $\pi_1(X)$ acts nilpotently on $H_*(\tilde{X})$.*

Proof. Let C be an SFP complex over $Z\pi_1(X)$ that is chain equivalent to $\text{CS}(\tilde{X})$. Since π' is of finite index in $\pi_1(X)$, it follows that C is SFP over $Z\pi'$. Since π' is a finite p group from Proposition 2.8, we see that $H_*(\tilde{X}) \simeq H_*(C)$ is a nilpotent π'-module. $\square$

Lemma 2.11

Let E be a subset of π generating π as a group. For any integer $k \geq 1$ let K_k be the subset of $Z\pi$ given by $D_1 = \{x - 1|x \in E\}$ and $K_k = \{\sigma_1 \cdots v_k|v_i \in D_1\}$ for $k \geq 2$. Then $(I\pi)^k$ is generated by D_k over $Z\pi$.

Proof. For $k = 1$, this is well known. Let $k \geq 1$ and assume it is valid for k. Then clearly $(I\pi)^{k+1}$ is generated by elements of the form $\gamma a(1 - x)$ with $\gamma \in K_k$, $a \in Z\pi$, and $x \in E$. Since $(I\pi)^k$ is a two-sided ideal of $Z\pi$, we have $\gamma a \in (I\pi)^k$. Hence $\gamma a = \Sigma \lambda_i \theta_i$ with $\lambda_i \in Z\pi$ and $\theta_i \in D_k$. Thus $\gamma a(1 - x) = \Sigma \lambda_i \theta_i(1 - x)$ and $\theta_i(1 - x) \in D_{k+1}$. $\square$

Proposition 2.12

Let S, T be subgroups of π such that $S \cup T$ generates π as a group. Assume further $st = ts$ in π for any $s \in S$ and $t \in T$. Let $M \in Z\pi$-mod be such that the actions of S and T on M are nilpotent. Then M is a nilpotent π-module.

Proof. Let $D_1 = \{x - 1 | x \in S \cup T\}$ and $D_k = \{v_1 \cdots v_k | v_i \in D_1\}$ for $k \geq 2$. Let $S_1 = \{s - 1 | s \in S\}$ and $T_1 = \{t - 1 | t \in T\}$. By Lemma 2.11 $(I\pi)^k$ is generated by D_k over $Z\pi$. Let m, n be integers greater than or equal to 1 satisfying $(IS)^m M = 0$ and $(IT)^n M = 0$. Since $st = ts$ for all $s \in S$ and $t \in T$, we see that elements of S_1 commute with elements of T_1. Hence any element of D_{m+n} will be of the form $u_1 \cdots u_k \cdot v_1 \cdots v_\ell$ with $u_i \in S_1$ for $1 \leq i \leq k$, $v_j \in T_1$ for $1 \leq j \leq \ell$, and $k + \ell = m + n$. Either $k \geq m$ or $\ell \geq n$. It follows that $(I\pi)^{m+n} M = 0$. $\square$

Corollary 2.13

Let C be an SFP complex over $Z\pi$ where π is a finite nilpotent group. Then $H_*(C)$ is a nilpotent π-module.

Proof. Since π is a finite nilpotent group, π is the direct product of its Sylow subgroups. Corollary 2.13 is now immediate from Proposition 2.8, Lemmas 2.9 and 2.11, and Proposition 2.12. $\square$

Corollary 2.14

Let X be a space with $\pi_1(X)$ finite and nilpotent. Suppose $CS(\tilde{X})$ is chain equivalent to an SFP complex over $Z\pi_1(X)$. Then X is nilpotent.

Proof. Immediate consequence of Corollary 2.13. $\square$

Remark 2.15

As observed in [14] X is a space of type FP and $\pi_1(X)$ is finitely presented, then X is finitely dominated. Accepting this result we get the following as an immediate consequence of Proposition 2.5 and Corollary 2.14.

Theorem 2.16

Let X be a space with $\pi_1(X)$ finite and nilpotent. Then X is finitely dominated, nilpotent if and only if $CS(\tilde{X})$ is chain equivalent to an SFP complex over $Z\pi_1(X)$.

3 WALL INVARIANT OF SFP COMPLEXES

Definition 3.1

Let C be an FP-complex over a ring R. The element $\Sigma(-1)^i[C_i]$ in $\tilde{K}_0(R)$ will be referred to as the (reduced) Wall invariant of C and will be denoted by $\tilde{w}(C)$. We can similarly define the unreduced Wall invariant $w(C)$ in $K_0(R)$.

For any abelian group π the integral closure of $Z\pi$ in $Q\pi$ will be denoted by $\overline{Z\pi}$. Let $j: Z\pi \to \overline{Z\pi}$ denote the inclusion. The kernel of $j_*: \tilde{K}_0(Z\pi) \to \tilde{K}_0(\overline{Z\pi})$ is denoted by $D(Z\pi)$. When π is a cyclic p group (namely a cyclic group of order p^k) it is known that $D(Z\pi)$ is a p group [2].

When π is cyclic of order p, $\overline{Z\pi} \simeq Z[\omega] \times Z$ where $\omega = \exp(2\pi\sqrt{-1}/p)$. From Corollary 5.13 in Chapter 5 we see that $D(Z\pi) = 0$ in this case.

Proposition 3.2

Let π be a cyclic p-group and C an SFP-complex over $Z\pi$. Then $\tilde{w}(C) \in D(Z\pi)$.

In particular if π is of order p, then $\tilde{w}(C) = 0$.

Proof. Let π be cyclic of order p^k. Then $\overline{Z\pi} = \prod_{j=0}^k B_j$ where $B_j = Z[\xi_j]$ with $\xi_j = \exp(2\pi\sqrt{-1}/p^j)$. In particular $B_0 = Z$. Let $\gamma(j): Z\pi \to Z[\xi_j]$ be the obvious ring homomorphism carrying the generator x of π to ξ_j. We have $\tilde{K}_0(B_0) = 0$. We will show that $\gamma(j)_*(\tilde{w}(C)) = 0$ in $\tilde{K}_0(B_j)$ for $j > 0$. The Cartan map $c: \tilde{K}_0(B_j) \to \tilde{G}_0(B_j)$ is an isomorphism since B_j is a Dedekind domain (Corollary 5.18, Chapter 3). It suffices to show that $c\gamma(j)_*(\tilde{w}(C)) = 0$. The element $c\gamma(j)_*(\tilde{w}(C))$ is equal to $\Sigma(-1)^i[H_i(B_j \otimes_\pi C)]$ in $\tilde{G}_0(B_j)$. Since $|\pi|\overline{Z\pi} \subset Z\pi$, from the homology exact sequence of

$$0 \longrightarrow Z\pi \otimes_\pi C \xrightarrow{j \otimes 1} \overline{Z\pi} \otimes_\pi C \longrightarrow (\overline{Z\pi}/Z\pi) \otimes_\pi C \longrightarrow 0,$$

we see that $j_*: H_*(C) \to H_*(\overline{Z\pi} \otimes_\pi C)$ is a $\Gamma(p)$ isomorphism, namely ker j_* and coker j_* are torsion abelian groups having only p-primary torsion. Since C is an SFP complex over $Z\pi$ and $|\pi| = p^k$, the map $H_*(C) \to H_*(Z \otimes_\pi C)$ is also a $\Gamma(p)$ isomorphism. From $\overline{Z\pi} = Z \times \prod_{j=1}^k B_j$ we get $H_*(\overline{Z\pi} \otimes_\pi C) = H_*(Z \otimes_\pi C) \oplus \oplus_{j=1}^k H_*(B_j \otimes_\pi C)$. It follows that $H_*(B_j \otimes_\pi C)$ is a finite

p-torsion abelian group for $j \geq 1$. Using the epimorphism $\gamma(j): Z\pi \to B_j$ we also get a left π action on B_j and hence a left π action on $H_*(B_j \otimes_\pi C)$. Since a p group acts only in a nilpotent way on a finite p-torsion abelian group, we see that $H_*(B_j \otimes_\pi C)$ is a nilpotent π-module. Hence the left B_j-module $H_i(B_j \otimes_\pi C)$ has a finite filtration such that on the graded associated the action of π is trivial. In particular $(1 - \xi_j)\mathrm{gr}\, H_i(B_j \otimes_\pi C) = 0$. Since $B_j/(1 - \xi_j)B_j \simeq Z/p$ with trivial π action, we see that $\mathrm{gr}\, H_i(B_j \otimes_\pi C)$ is a finite-dimensional Z/p vector space with trivial π action (and hence B_j acting by $\xi_j \cdot a = a \;\forall\; a \in Z/p$). From the exact sequence $0 \to B_j \xrightarrow{(1-\xi_j)} B_j \to Z/p \to 0$ we get $[Z/p] = [B_j] - [B_j] = 0$ in $\tilde{G}_0(B_j)$. Hence $c\gamma(j)_*(\tilde{w}(C)) = 0$ and hence $\gamma(j)_*(\tilde{w}(C)) = 0$ for $j \geq 1$. This really means the element $w(C) \in K_0(Z\pi)$ goes to $\sum_{j=0}^k n_j[B_j] \in K_0(\overline{Z\pi}) \simeq \oplus_{j=0}^k K_0(B_j)$ under j_* with $n_j \in Z$. The commutativity of the diagram

$$
\begin{array}{ccccc}
K_0(Z\pi) & \longrightarrow & K_0(\overline{Z\pi}) & \longrightarrow & K_0(B_j) \\
{\scriptstyle \varepsilon_*}\downarrow & & & & \downarrow \\
Z \simeq K_0(Z) & & = & = & K_0(Z)
\end{array}
$$

shows that $n_j = \varepsilon_*(w(C))$ for all $j \geq 0$. Thus if $\varepsilon_*(w(C)) = n$, then $j_*(w(C)) = n\sum_{j=0}^k[B_j] = n[\overline{Z\pi}]$. Hence $j_*(\tilde{w}(C)) = 0 \in \tilde{K}_0(\overline{Z\pi})$. $\square$

Theorem 3.3.

Let π be a finite p group and C an SFP complex over $Z\pi$. Then $\tilde{w}(C)$ is an element in the p-primary torsion subgroup of $\tilde{K}_0(Z\pi)$.

For proving this theorem we need the following result of Swan ([22], Corollary 9.4).

Proposition 3.4

Let π be a group of order n and $x \in \tilde{K}_0(Z\pi)$. Suppose $i^*x = 0$ in $\tilde{K}_0(Z\pi')$ for all $i: \pi' \twoheadrightarrow \pi$ with π' cyclic. Then $nx = 0$.

Proof of Theorem 3.3. Let π' be a cyclic subgroup of π. As observed already (Lemma 2.9), C is an SFP complex over $Z\pi'$. Let $\tilde{w}(C) \notin \tilde{K}_0(Z\pi)$ be the Wall invariant of C regarded as a complex over $Z\pi$. Then the Wall invariant of C regarded as a complex over $Z\pi'$ is $i^*(\tilde{w}(C)) \in \tilde{K}_0(Z\pi')$ where $i: \pi' \to \pi$ is the inclusion. From Proposition 3.2 and the fact that $D(Z\pi')$ is a p-torsion abelian group we see that $i^*(p^{\ell(\pi')}\tilde{w}(C)) = 0$ for some integer $\ell(\pi')$ depending on π'. Since there are only finitely many subgroups of π, there exists an integer ℓ with $i^*(p^\ell \tilde{w}(C)) = 0$ for every $i = \pi' \to \pi$ with π' cyclic. If $|\pi| = p^k$, from Proposition 3.4 we get $p^{k+\ell}\tilde{w}(C) = 0$. □

As an immediate consequence of Proposition 2.5, Proposition 3.2, and Theorem 3.3 we get the following:

Theorem 3.5

Let X be a finitely dominated nilpotent space with $\pi_1(X) = \pi$ a cyclic p group. Then $\tilde{w}(X) \in D(Z\pi)$. In particular if $|\pi| = p$, then $\tilde{w}(X) = 0$.

Theorem 3.6

Let X be a finitely dominated nilpotent space with $\pi(X)$ a finite p group. Then $\tilde{w}(X)$ is an element in the p-torsion subgroup of $\tilde{K}_0(Z\pi_1(X))$.

Remark 3.7

An example of a finitely dominated nilpotent space X with $\pi_1(X) = Q(8)$ the quaternion group $\{\pm 1, \pm i, \pm j, \pm k\}$ of order 8 having $\tilde{w}(X) \neq 0$ is given in Mislin's paper [14]. A full discussion of this example will take us too far from the scope of the present book. The interested reader may refer to Mislin's paper.

Our next aim is to show that $\tilde{w}(C) \in D(Z\pi)$ for any SFP complex C over $Z\pi$ where π is any finite abelian group. As a first step we prove:

Proposition 3.8

Let π be a finite cyclic group and C an SFP complex over $Z\pi$. Then $\tilde{w}(C) \in D(Z\pi)$.

Proof. Let $|\pi| = n = p_1^{k_1} \cdots p_r^{k_r}$ with $k_i \geq 1$ and $p_1, \ldots, p_r$ distinct primes. We will prove the proposition by induction on the number r of distinct primes occurring in the factorization of $|\pi|$. When $r = 1$, this is exactly Proposition 3.2.

Now assume $r > 1$. Let x be a chosen generator of π. Let $n_i = n/p_i^{k_i}$. Then $y_i = x^{n_i}$ generates a subgroup π_i of order $p_i^{k_i}$ and $\pi \simeq \pi_1 \times \cdots \times \pi_r$ with projection $\eta_i: \pi \to \pi_i$ carrying x to y_i. The integral closure $\overline{Z\pi}$ of $Z\pi$ in $Q\pi$ is $\prod_{d/n} A_d$ where $A_d = Z[\omega_d]$ with $\omega_d = \exp(2\pi\sqrt{-1}/d)$ and d runs over all the divisors of n. Let $\gamma(d): Z\pi \to Z[\omega_d]$ be the obvious ring homomorphism carrying x to ω_d. To prove Proposition 3.8 we have only to show that $\gamma(d)_*(\tilde{w}(C)) = 0$ for all divisors d of n.

Case I.

Suppose d does not involve at least one of the primes p_i. Without loss of generality we may assume p_r is not involved. Then $d = p_1^{\ell_1} \cdots p_{r-1}^{\ell_{r-1}}$ with $0 \le \ell_i \le k_i$ for $1 \le i \le r - 1$. Let $\varphi\colon \pi \to \pi_1 \times \cdots \times \pi_{r-1}$ denote the projection and $u = \varphi(x)$. Writing π' for $\pi_1 \times \cdots \times \pi_{r-1}$ let $\gamma'(d)\colon Z\pi' \to A_d$ be the obvious map carrying u to ω_d. Then $\gamma(d) = \gamma'(d) \circ \varphi$ where $\varphi\colon Z\pi \to Z\pi'$ is induced by $\varphi\colon \pi \to \pi'$. Hence $\gamma(d)_*(\tilde{w}(C)) = \gamma'(d)_*(\varphi_*(\tilde{w}(C)))$. But $\varphi_*(\tilde{w}(C)) = \tilde{w}(Z\pi' \otimes_\pi C)$. From Proposition 2.7 we see that $Z\pi' \otimes_\pi C$ is an SFP complex over $Z\pi'$. Since $|\pi'| = p_1^{k_1} \cdots p_{r-1}^{k_{r-1}}$, with only $(r - 1)$ distinct primes occurring, by the inductive assumption we see that $\gamma'(d)_*(\tilde{w}(Z\pi' \otimes_\pi C)) = 0$. Hence $\gamma(d)_*(\tilde{w}(C)) = 0$.

We have to deal with the case when d involves all the primes $p_1, \ldots, p_r$, say $d = p_1^{\ell_1} \cdots p_r^{\ell_n}$ with $1 \le \ell_i \le k_i$ for $1 \le i \le r$.

We need certain observations. Since $|\pi|\overline{Z\pi} \subset Z\pi$ from the homology exact sequence of

$$0 \to Z\pi \otimes_\pi C \to \overline{Z\pi} \otimes_\pi C \to (\overline{Z\pi}/Z\pi) \otimes_\pi C \to 0,$$

we see that $j_*\colon H_*(C) \to H_*(\overline{Z\pi} \otimes_\pi C) = H_*(Z \otimes_\pi C) \oplus \oplus_{d>1} H_*(A_d \otimes_\pi C)$ is a $\Gamma(n)$ isomorphism. Since C is SFP, $H_*(C) \to H_*(Z \otimes_\pi C)$ is a $\Gamma(n)$ isomorphism. It now follows (because π is finite, hence $Z\pi$ noetherian) that $H_*(A_d \otimes_\pi C)$ for $d > 1$ is a finite abelian group with no q torsion for any $q \in \Gamma(n)$.

Case II.

$d = p_1^{\ell_1} \cdots p_r^{\ell_r}$ with $1 \le \ell_i \le k_i$ for $1 \le i \le r$. Let L_i be the p_i torsion of $H_*(A_d \otimes_\pi C)$. From the preceding comments we have $H_*(A_d \otimes_\pi C) = L_1 \oplus \cdots \oplus L_r$. Since any automorphism of a group carries p_i torsion to itself, it follows that each L_i is a π-submodule of $H_*(A_d \otimes_\pi C)$. Now choose a prime q different from every one of the p_i's. Let τ be a cyclic group of order q and b a chosen generator from τ. Let $H = \pi \times \tau$. Then H is cyclic of order nq with $u = (x, b)$ as a generator. Let $\gamma'(dq)\colon ZH \to A_{dq}$ be the obvious map carrying u to $\omega_{dq} = \exp(2\pi\sqrt{-1}/dq)$. Using the projection $H \xrightarrow{\varphi} \pi$ we regard $Z\pi$ as a left ZH-module and using $\gamma'(dq)\colon ZH \to A_{dq}$ we regard A_{dq} as a right ZH-module. It can be checked that there exist right $Z\pi$ isomorphisms $A_{dq} \otimes_H Z\pi \xrightarrow{g} A_d$, $A_d \xrightarrow{h} A_{dq} \otimes_H Z\pi$ satisfying $g(\omega_{dq}^i \otimes_H \alpha)\omega_d^i \cdot \alpha$; $h(\omega_d^i) = \omega_{dq}^i \otimes_H 1$, for every integer $i \ge 0$ and $\alpha \in \pi$. It follows that $(A_{dq} \otimes_H Z\pi) \otimes_\pi C \simeq A_d \otimes_\pi C$ as complexes over Z. If τ acts trivially on C, from the preceding isomorphism we get $A_{dq} \otimes_H C \simeq A_d \otimes_\pi C$ as complexes over Z and hence

$$H_*(A_{dq} \otimes_H C) \simeq H_*(A_d \otimes_\pi C) = L_1 \oplus \cdots \oplus L_r.$$

Using $\gamma'(dq)$: $ZH \to A_{dq}$ and $\gamma(d)$: $Z\pi \to A_d$, we regard A_{dq} as a left H-module and A_d as a left π-module. We make τ act trivially on the left on A_d and convert A_d into a left H-module. The isomorphism g: $A_{dq} \otimes_H Z\pi \to A_d$ is then a left $Z\pi$ isomorphism. Actually

$$g\left(u \cdot \omega_{dq}^i \otimes_H \alpha\right) = g\left(\omega_{dq}^{i+1} \otimes_H \alpha\right) = \omega_d^{i+1} \cdot \alpha$$

$$= x \cdot \omega_d^i \alpha$$

$$= u \cdot \omega_d^i \alpha$$

since τ acts trivially. Hence $H_*(A_{dq} \otimes_H C) \simeq L_1 \oplus \cdots \oplus L_r$ with trivial τ action. Since L_i is a p_i group, we see that π_i acts nilpotently on L_i. Since τ acts trivially, it follows that $\pi_i \times \tau$ acts nilpotently on L_i. Hence there exists a finite filtration $L_i = F_0 \supset F_1 \supset \cdots \supset F_k = 0$ by $\pi_i \times \tau$-submodules of L_i with $F_\mu/F_{\mu+1}$ a trivial $\pi_i \times \tau$-module, $0 \leq \mu \leq k - 1$. Now, $u^{n_i} = (x^{n_i}, b^{n_i})$ generates $\pi_i \times \tau$ [since $(n_i, q) = 1$] and under $\gamma'(d_q)$ the element u^{n_i} is mapped to $\omega_{dq}^{n_i}$. Since n_i involves one prime less than d, namely p_i is not involved in n_i, we see that $\omega_{dq}^{n_i}$ is a root of unity of mixed order. Hence $1 - \omega_{dq}^{n_i}$ is a unit in $Z[\omega_{dq}^{n_i}]$. From $(1 - \omega_{dq}^{n_i})F_\mu/F_{\mu+1} = 0$ we get $F_\mu/F_{\mu+1} = 0$ for $0 \leq \mu \leq k - 1$. Hence $L_i = 0$. It follows that $H_*(A_d \otimes_\pi C) = 0$. Now, if c: $\tilde{K}_0(A_d) \to \tilde{G}_0(A_d)$ denotes the Cartan isomorphism, we have $C(\gamma(d)_*)(\tilde{w}(C)) = \Sigma(-1)^\nu[H_\nu(A_d \otimes_\pi C)] = 0$ in $\tilde{G}_0(A_d)$. Hence $\gamma(d)_*(\tilde{w}(C)) = 0$. This proves Proposition 3.8. $\square$

Theorem 3.9

Let π be a finite abelian group and C an SFP complex over $Z\pi$. Then $\tilde{w}(C) \in D(Z\pi)$.

Proof. Let $|\pi| = m$. If $\chi_1, \ldots, \chi_k$ are the irreducible rational characters of π, then the integral closure $\overline{Z\pi}$ of $Z\pi$ in $Q\pi$ is shown to be $\prod_{i=1}^k A_i$ where $A_i = Z[\omega_{m_i}]$ with m_i equal to order χ_i, namely χ_i arises from a homomorphism ρ_i of π to the complex roots of unity with image a group of order m_i. If $\pi_i = \ker \rho_i$, then π/π_i is known to be cyclic of order m_i. Moreover the map $Z\pi \to A_i$ is known to factor as $Z\pi \xrightarrow{\varphi} Z(\pi/\pi_i) \xrightarrow{\gamma(m_i)} A_i$ where φ is induced by the quotient map $\pi \to \pi/\pi_i$ and $\gamma(m_i)$: $Z(\pi/\pi_i) \to Z[\omega_{m_i}] = A_i$ is the obvious map carrying a generator of π/π_i to ω_{m_i}. (See pages 54 and 55 of [1].) Writing π' for π/π_i, we see from Proposition 2.7 that $Z\pi' \otimes_\pi C$ is SFP over π'. By Proposition 3.8 we have $\gamma(m_i)_*(\tilde{w}(Z\pi' \otimes_\pi C)) = 0$. But under the map $Z\pi \to A_i$ the element $\tilde{w}(C)$ in $\tilde{K}_0(Z\pi)$ gets mapped to $(\gamma(m_i) \circ \varphi)_*(\tilde{w}(C)) = \gamma(m_i)_*(\tilde{w}(Z\pi' \otimes_\pi C)) = 0$. This completes the proof of Theorem 3.9. $\square$

As an immediate consequence we get the following theorem.

Theorem 3.10

Let X be a finitely dominated nilpotent space with $\pi_1(X)$ finite abelian. Then $\tilde{w}(X) \in D(Z\pi_1(X))$.

For further results on the Wall obstruction of nilpotent spaces the reader can refer to [15].

References

1. H. Bass. *Algebraic K-Theory*. W. A. Benjamin, 1968.
2. H. Bass and M. P. Murthy. Grothendieck groups and Picard groups of abelian group rings. *Ann. of Math.* **86** (1967), pp. 16–73.
3. K. S. Brown and E. Dror. The Artin–Rees property and homology. *Israel J. Math.* **22** (1975), pp. 93–109.
4. H. Cartan and S. Eilenberg. *Homological Algebra*. Princeton University Press, 1973.
5. C. W. Curtis and I. Reiner. *Representation Theory of Finite Groups and Associative Algebras*. Interscience Publishers, 1966.
6. E. Dror. A generalisation of the Whitehead theorem. *Lecture Notes in Math.* **249**. Springer, 1971, pp. 13–22.
7. S. M. Gersten. A product formula for Wall's obstruction. *Amer. J. Math.* **88** (1966), pp. 337–346.
8. P. Hilton. Nilpotent actions on nilpotent groups. *Lecture Notes in Math.* **450**. Springer, 1975, pp. 174–197.
9. P. Hilton, G. Mislin, and J. Roitberg. *Localization of Nilpotent Groups and Spaces*. *Math. Studies* **15**. North-Holland, 1975.
10. V. J. Lal. The Wall obstruction of a fibration, *Invent. Math.* **6** (1968), pp. 67–77.
11. A. T. Lundell and S. Weingram. *The Topology of CW-Complexes*. Van Nostrand Reinhold, 1969.
12. J. W. Milnor. The geometric realization of a semi-simplicial complex. *Ann. of Math.* **65** (1957), pp. 357–362.
13. G. Mislin. Wall's obstruction for nilpotent spaces. *Topology* **14** (1975), pp. 311–317.
14. G. Mislin. Finitely dominated nilpotent spaces. *Ann. of Math.* **103** (1976), pp. 547–556.
15. G. Mislin and K. Varadarajan. The finiteness obstructions for nilpotent spaces lie in $D(Z\pi)$. *Invent. Math.* **53** (1979), pp. 185–191.
16. I. Reiner. Integral representations of cyclic groups of prime order. *Proc. Amer. Math. Soc.* **8** (1957), pp. 142–146.

17. D. S. Rim. Modules over finite groups. *Ann. of Math.* **69** (1959), pp. 700–712.

18. J. Roitberg. Note on nilpotent spaces and localization. *Math. Z.* **137** (1974), pp. 67–74.

19. E. H. Spanier. *Algebraic Topology.* McGraw-Hill, 1966.

20. J. R. Stallings. Homology and central series of groups. *J. Algebra* **2** (1965), pp. 170–181.

21. N. Steenrod. *The Topology of Fibre Bundles.* Princeton University Press, 1951.

22. R. G. Swan. Induced representations of projective modules. *Ann. of Math.* **71** (1960), pp. 552–578.

23. K. Varadarajan. Finiteness obstruction for nilpotent spaces. *J. Pure Appl. Algebra* **12** (1978), pp. 137–146.

24. C. T. C. Wall. Finiteness conditions for CW-complexes. *Ann. of Math.* **81** (1965), pp. 56–69.

25. C. T. C. Wall. Finiteness conditions for CW-complexes. II. *Proc. Roy. Soc. London Ser. A* **295** (1966), pp. 129–139.

26. G. W. Whitehead. *Elements of Homotopy Theory. Graduate Texts in Math.* **61**. Springer, 1978.

27. J. H. C. Whitehead. Combinatorial homotopy. I. *Bull. Amer. Math. Soc.* **55** (1949), pp. 213–245.

28. J. H. C. Whitehead. Combinatorial homotopy. II. *Bull. Amer. Math. Soc.* **55** (1949), pp. 453–496.

29. O. Zariski and P. Samuel. *Commutative Algebra*, Vol. I. Van Nostrand Reinhold, 1958.

Index